THE DICTIONARY OF ENVIRONMENTAL SCIENCE AND ENGINEERING

The Dictionary of Environmental Science and Engineering is an essential reference book for students and professionals in such diverse disciplines as geography, earth sciences, biology, engineering and law. Expanded from its original publication, this comprehensive and concise guide defines the scientific and the technical terminology that surrounds this high profile topic.

The dictionary includes:

- over 800 definitions, from Absorption to Zeno's Paradox
- an extensive list of useful acronyms and abbreviations
- coverage of the latest emerging ideas and terminologies within the discipline
- a clear, no-nonsense approach to a complex field

James R. Pfafflin is a professional engineer with degrees from Indiana State University, Johns Hopkins University and the University of Windsor. He is a member of the Commissioned Reserve of the US Public Health Service.

Edward N. Ziegler is Associate Professor of Chemical and Biological Engineering at Polytechnic University. He is a consultant to both private industry and government.

Joseph M. Lynch is a professional engineer and a founding partner of the environmental engineering consulting firm Mayo Lynch Associates.

The Dictionary of Environmental Science and Engineering

Second edition

**James R. Pfafflin,
Edward N. Ziegler,
and Joseph M. Lynch**

Routledge
Taylor & Francis Group

NEW YORK AND LONDON

First published 1996 by Gordon and Breach Publishers
This edition published 2008
by Routledge
270 Madison Avenue, New York, NY 10016

Simultaneously published in the UK
by Routledge
2 Park Square, Milton Park, Abingdon, Oxon OX14 4RN

Routledge is an imprint of the Taylor & Francis Group, an informa business

© 2008 James R. Pfafflin

Typeset in Bembo and Gill Sans by
Taylor & Francis Books
Printed and bound in Great Britain by
Antony Rowe Ltd, Chippenham, Wiltshire

Library of Congress Cataloging in Publication Data
A catalog record for this book has been requested

British Library Cataloguing in Publication Data
A catalogue record for this book is available from the British Library

ISBN 978-0-415-77194-8 (hbk)
ISBN 978-0-415-77195-5 (pbk)
ISBN 978-0-203-93405-0 (ebk)

This work is dedicated to the late Mr. Leonard A. Murphy
and the late Mrs. Honora Shields Murphy

Contents

Preface

The broad field now known as environmental science and engineering is truly interdisciplinary. The many subjects which comprise this major study area have unique languages which are often understood only by specialists. There is a need for a source which workers in various areas can consult for guidance as to meanings and concepts. Attributed to George Bernard Shaw is the observation that England and America are two great nations separated by a common language. This problem has arisen in preparation of this work and it was decided to use American spelling throughout.

The beginnings of environmental science and engineering were based in public health practice. Environmental efforts are now much broader. Thus, the work presented here is not complete. An assertion of completeness would be a claim that one can define the nature and limits of components of environmental science and engineering. This is not possible for such a dynamic situation. One can but present terms which have appeared in pertinent literature.

Much significant legislation, good and bad, was passed before the advent of the Environmental Movement and an even greater amount since that time. Acronyms have come into wide use, both legal and scientific, and a listing of these should be helpful. The expected audience will be scientists, engineers, public officials, administrators, managers, attorneys, students, journalists and interested laymen.

A

A (AEROBACTER) AEROGENES
A soil dwelling microorganism which gives a false positive test for the presence of coliforms. see **Aerobes**. see *E. coli*

ABBAN
see **Endogenous respiration**

ABSOLUTE LIMEN
The minimum amount of stimulus energy required to elicit a response.

ABSORBANCE
see **Optical density**

ABSORPTION
A diffusion process which involves transfer of material from the gas phase to the liquid phase due to the concentration gradient. see **Diffusion**

ACETALDEHYDE
This compound can be a contributor to a number of pathologies ranging from cancer to asthma. Origins are bacteria in the oral cavity and intestinal tract due to industrial and environmental pollution. It is found in cigarette smoke.

ACETIC ACID (HAc)
The most common aliphatic acid in sewage, it is often written as CH_3COOH. It is a weakly ionized acid and is thought to be significant

in early warning of problems with anaerobic digestion of sewage sludge.
see **Volatile acids**

ACID DEPOSITION

Transfer of acidic substances from the atmosphere to the earth's surface
by both wet and dry deposition processes.

ACIDITY

The equivalent sum of acids that are titratable with strong base.

ACID MINE DRAINAGE

Sulfuric acid and iron compounds result from reaction of water and air
with sulfur bearing minerals in active and abandoned coal mines. pH
of streams and other receiving waters is lowered as a result. Control
measures include sealing off underground mines, restoration of surface
strip mines and collection and treatment of drainage.

ACID PRECIPITATION

Rain, sleet and snow with pH of less than 5.6. This is the equilibrium
pH for atmospheric CO_2 (0.03%) in contact with water.

ACID RAIN (ACID SNOW)

Precipitation with pH less than 5.6. Pure water, in equilibrium with
carbon dioxide in the atmosphere, has a pH of about 5.6. This value is
taken as zero for acid rain study. When an acid particle in the atmosphere
is scrubbed out, the pH of the precipitation will fall. The primary sources
of the acid particles are combustion products of sulfur bearing fuels.

ACTINIDES

Elements with atomic numbers from 89 to 103. All are produced by
radioactive decay or artificially in reactors.

ACTINOMYCETES

Naturally branching filamentous spore forming organisms resembling
fungi. These are not true bacteria. Some Actinomycetes of the genus
Streptomyces are commercially important sources of antibiotics.

ACTIVATED CARBON

A porous bead or powder resulting from charring of a carbon source
to drive off volatile matter, this has great surface area per unit volume.

It is widely used for removal of tastes, odors and undesirable substances from air and water.

ACTIVATED SLUDGE.

A biological wastewater treatment process in which voraciously feeding microorganisms are suspended in the flowing water. The microorganisms utilize the waste as food. This is a strictly aerobic process. Aeration provides the necessary dissolved oxygen and mixing. Wastes enter the aeration tank after mixture with returned sludge. Returned sludge provides food to keep the microorganisms in an active growth phase. The microbial mass (mixed liquor suspended solids) to which the waste has been transferred is settled out in a secondary (humus) tank. A part of the settled sludge is recirculated and the rest goes to a sludge digester. The process requires careful operational control. Modifications of the basic process have usually resulted from specific operating conditions. Environmental factors governing the process are temperature, pH and oxidation-reduction potential.

ACTIVATION PRODUCTS

Man-made radionuclides. Examples are cobalt-60, zinc-65, calcium-45 and phosphorous-32.

ACTIVE METABOLIC RATE

The maximum level of oxygen consumption of an actively swimming fish.

ACTIVITY

Effective concentration. The actual concentration of a substance in solution, expressed as mol/L, times the activity coefficient gamma. Gamma is a function of the concentrations of other substances in the solution. see **Ionic strength**

ACUTE TOXICITY

Severe biological effects or death caused by a single exposure to a substance.

ADENOSINE TRIPHOSPHATE (ATP)

Chemical energy is stored in the bonds of ATP. A considerable amount of this energy is released when certain bonds are broken and this is the energy which operates all autotrophic and heterotrophic microbial cells. see **Autotrophic**. see **Heterotrophic**

ADIABATIC
No net gain or loss of heat.

ADIABATIC LAPSE RATE
Air becomes cooler with increasing distance from the earth's surface. A normal adiabatic lapse rate is 1°C per 100 meters. A rate greater than this is a superadiabatic lapse rate and is associated with good mixing. A rate less than the normal is a weak adiabatic lapse rate and is associated with light winds. A temperature inversion occurs when temperature increases with height. This is a stable condition. see **Atmospheric inversion**

ADSORBATE
see **Adsorption**

ADSORBENT
see **Adsorption**

ADSORPTION
The process in which a gas or liquid accumulates on the surface of a solid or liquid to form a molecular film. The surface on which material forms is the adsorbent. The material forming on the surface is the adsorbate.

ADULTERANT
A substance that, by law, should not be in a food or pesticide.

ADVANCED OXIDATION TECHNOLOGY (AOT)
Included are ultraviolet radiation, ozonation and hydrogen peroxide addition. These are used to remove organic contaminants and inactivate microorganisms. Ultraviolet radiation has some drawbacks, primarily hardware, but is relatively inexpensive. Ozone is quite effective for microbial inactivation and is somewhat effective in breaking down organic compounds. It must be generated on site. Hydrogen peroxide can be used to enhance treatment with ultraviolet radiation and ozone. see **Ozone**. see **Ultraviolet radiation**

AERATED LAGOONS
These are similar to activated sludge but without recirculation. Air diffusers or mechanical aerators provide dissolved oxygen and mixing. Deposited solids undergo anaerobic decomposition.

AERATION ZONE
The soil zone above the water table in contact with the atmosphere.

AEROBES
Microorganisms which require dissolved molecular oxygen in carrying out metabolism of food.

AEROBIC BIOLOGICAL TREATMENT PROCESSES
These include fixed film or stationary contact (trickling filter) and suspended contact (activated sludge).

AERODYNAMIC DIAMETER
The diameter of a spherical particle having a density of 1 gm/cm^3 (water) which settles at the same rate as the particle under consideration.

AEROMETRIC NETWORKS
Set up by cities and towns, these have gathered and analyzed data on air pollutants for many years. Informing the public on air quality in a simple and understandable manner is necessary and led the EPA to develop the Air Quality Index. see **Air Quality Index**

AEROSOL
A system of tiny particles suspended in a gas. Any substance (except pure water) which exists as a liquid or solid in the atmosphere under normal conditions and is of microscopic or submicroscopic size but larger than molecular dimensions.

AFTERBURNING
Injection of air downstream from the exhaust valve of an internal combustion engine for conversion of unburned hydrocarbons and carbon monoxide to carbon dioxide and water vapor. NO_x is not removed by an afterburner and, in fact, more is formed. This is a disadvantage.

AGENDA 21
Adopted at the Conference on Environment and Development (Earth Summit) in Rio, this is a wide ranging blueprint for action worldwide to achieve sustainable development.

AGENT ORANGE
A chemical defoliant used during the Vietnam conflict. Civilians as well as combatants on both sides were exposed and cancers, deformities,

stillbirths and blindness are attributed to these exposures. It is thought that the adverse effects of Agent Orange are due to dioxins present as impurities. see **Dioxin**

AGGREGATE RISK

The product of the individual risk and the population under consideration. see **Unit risk value**

AHERA

see **Asbestos Hazard Emergency Response Act**

A-HORIZON

see **Soil**

AIRBORNE INFECTIONS

Some organisms cannot survive long outside a host while others can remain viable for long periods. Some of the major diseases classified as human airborne infections are chicken pox, common cold, diphtheria, german measles, hemolytic streptococci, influenza, measles, meningococcus meningitis, mumps, mycoplasma pneumonia, pneumonococcus pneumonia, poliomyelitis, psittacosis, rheumatic fever, scarlet fever, smallpox, systemic mycosis and tuberculosis.

AIR POLLUTION

The presence in the atmosphere of any substance (or combination of substances) that is detrimental to the health and welfare of humans or other living matter, offensive or objectionable to people or other living matter, either externally or internally, or which, by its presence, will affect adversely the welfare of people or other living matter, either directly or indirectly.

AIR QUALITY INDEX

Developed by the US EPA, this reports daily levels of ozone, particulate matter, carbon monoxide, sulfur dioxide and nitrogen dioxide on a color-coded scale of 0 to 500.

0–50	good (green)
51–100	moderate (yellow)
101–50	unhealthy for sensitive groups (orange)
151–200	unhealthy (red)

201–300   very unhealthy (purple)
301–500   hazardous (maroon)

AIRSHED MODEL
A mathematical model based on emissions, topography, meteorology and chemistry which can predict concentrations of primary and secondary pollutants at various locations in the model domain. These usually treat a small geographic area of the order of 100 km^2.

AIR STRIPPING
Nitrogen present in wastewaters is in the ammonia and organic forms. High nitrogen levels in the ammonia form can be removed by elevating the pH above 10 and agitating in a counter flow operation. VOCs can be removed by air stripping and it is not necessary to adjust the pH.

AL
see **Absolute limen**

ALAR
A substance used on apples, peanuts, cherries and other fruit to enhance color and firmness.

ALASKA LANDS ACT
see **Alaska National Interest Lands Conservation Act**

ALASKA NATIONAL INTEREST LANDS CONSERVATION ACT (US)
In 1978 the President, under authority of the Antiquities Act, designated 17 National Monuments, covering 56 million acres. An additional 50 million acres of Alaskan land were withdrawn for wildlife refuge under the Federal Land Policy and Management Act. In 1980 Congress passed the Alaska National Interest Lands Conservation Act, which ratified most of the previous withdrawals made under executive orders. see **Antiquities Act**. see **Federal Land Policy and Management Act**

ALBEDO
The portion of the solar radiation reaching the earth's surface which is reflected immediately back into the atmosphere. The average value is about 40% but varies locally with the nature of the surface on which the radiation impinges. see **Solar radiation**

ALCOHOLS

These are primary oxidation products of hydrocarbons and are classified as primary, secondary or tertiary, depending on where OH groups are attached to the molecule. Alcohols react with organic acids to form esters.

ALDEHYDES

One of a group of organic compounds which yield acids when oxidized and alcohols when reduced. These are products of incomplete combustion. All contain the carbonyl groups.

ALDRIN

see **Organochlorine insecticides**

ALFISOLS

see **Tropical soils**

ALGAE

These include all unicellular and multicellular microscopic plants which carry out true photosynthesis. Algae are autotrophic and utilize inorganic compounds for production of protoplasm. Carbon dioxide is the carbon source and ammonia, nitrite or nitrate are nitrogen sources. Oxygen is produced during growth. Eutrophocation can be characterized as explosive algal growth and chlorination can produce taste and odor problems in water supplies due to material released from dead algal cells. see **Autotrophic**. see **Eutrophication**

ALGAL BLOOM

Explosive increase in algae growth which can have adverse effects on water quality and fish life. see **Eutrophication**

ALIPHATIC COMPOUNDS

These contain only straight or branched carbon chains.

ALKALINITY

The ability of water to neutralize acids (resisting pH change). The carbonate–bicarbonate system is the major source of natural alkalinity. The dominant form of alkalinity is pH dependent. Alkalinity is expressed in eq/L or mg/L $CaCO_3$. Calcium carbonate has a molecular weight of 100 and this makes it convenient for expression. Alkalinity may also be reported as meq/L. 1 mol/L of calcium carbonate = 2 eq/L.

ALKANES

These are also called paraffins. Lower alkanes are colorless gases or liquids but higher alkanes (17 or more carbons in the chain) are solids. There are straight chain alkanes, branched alkanes and cycloalkanes. The simplest alkane is methane.

ALKENES

These compounds have a carbon-carbon double bond. The general formula is C_nH_{2n}. Because of the double bond, these are unsaturated.

ALKYL GROUP

This is produced when one hydrogen atom is removed from an alkane.

ALKYNES

Hydrocarbons which contain a triple bond. The general formula is C_nH_{2n-2}.

ALLOCHTHONOUS MATERIAL

That material which is brought from outside the system under consideration.

ALLOTROPES

Elements which can exist in more than one form. These are carbon, selenium, bismuth and sulfur. Graphite is an example.

ALPHA RAY (PARTICLE)

These are not true rays but are particles of matter. They are double charged ions of helium and are easily attenuated but have strong ionizing power.

ALVEOLI

Thin walled air sacs in the respiratory system.

AMINE

Alkyl derivatives of ammonia. These can occur as primary, secondary or tertiary, depending on the number of ammonia hydrogen atoms replaced. Tertiary amines combine with alkyl halides to form quaternary ammonium salts, which are used as disinfecting agents.

AMINO ACIDS
These act as building blocks from which proteins are constructed. Free amino acids are amphoteric (can function as acids or bases).

AMMONIA
Freshly polluted waters have nitrogen in the organic (protein) form and ammonia. Ammonia produced by bacterial action on urea and proteins can be used directly by plants to produce more protein. Any excess is oxidized by bacteria of the *Nitrosomonas* group to nitrite and then by the *Nitrobacter* group to nitrate. Ammonia, nitrite and nitrate are usually reported in terms of the nitrogen in the ion as NH_3–N, NO_2^-–N and NO_3^-–N.

AMMONIA SLIP
Ammonia (NH_3) is used to reduce NO_x emissions in the tail gas of industrial processes. The ammonia slip is the NH_3 converted to nitrogen and water in the exhaust stream of the stack.

AMPEROMETRIC
A current which is generated due to oxidation-reduction potential.

AMPHOTERIC
A substance which can function as an acid or base.

ANAEROBES
Microorganisms which carry out their life processes without using dissolved molecular oxygen for food metabolism.

ANAEROBIC
Absence of dissolved molecular oxygen. In wastewater and in rivers, anaerobic conditions develop when dissolved oxygen and nitrate as oxygen sources are exhausted. The oxygen source then is sulfate and objectionable conditions develop.

ANAEROBIC WASTEWATER TREATMENT
Anaerobic bacteria stabilize organic matter in the absence of free oxygen. One advantage is that much of the organic matter is converted to liquid or gas and much less sludge is produced.

ANNEX 21
see **Kyoto Protocol**

ANOXIA
The condition where oxygen does not reach body tissues or is not utilized by the tissues.

ANOXIC
No dissolved molecular oxygen is present.

ANTAGONISM
The process by which the toxic effects of a substance are reduced due to the presence of a second or more compounds. This is the reverse of potentiation or synergy. see **Potentiation**. see **Synergism**

ANTARCTIC TREATY
A 1959 agreement initially signed by 12 nations. It was agreed that the Antarctic would be nuclear free, demilitarized and devoted to research. There has since been pressure by some countries to initiate resource exploration and exploitation.

ANTHRAFILT
The name given to crushed coal used as a filter medium in water treatment.

ANTHROPOGENIC
Relating to the impact of people on Nature.

ANTIBIOTICS
Relatively complex chemical substances of microbial origin which display antimicrobial activity. Examples are penicillin and streptomycin.

ANTIMONY
A soft, silvery-white metal which occurs in Nature as the sulfide (Sb_2S_3). Physiological effects are similar to those of arsenic but vomiting and eye and mucous membrane irritation may be more severe.

ANTIQUITIES ACT (US)
The Secretary of the Interior is empowered to designate certain federally owned lands as National Monuments. These include historic landmarks, historic and prehistoric structures and other objects of historic or scientific interest. The Secretary can prohibit any activity that would affect the site adversely.

ANTISEPSIS
Destruction of microorganisms but not bacterial spores or living tissue. Although all microorganisms may not be killed, they will be reduced to a population level not considered harmful to health. Joseph Lister used this method in developing asepis in surgery. He was inspired by the use of phenol for control of sewage odors.

AONB
see **Area of Outstanding Natural Beauty**

AOT
see **Advanced Oxidation Technology**

AOX
see **Organochlorine**

AQUEDUCT
An artificial channel for carrying water.

AQUICLUDE
An impervious stratum (rock, clay, etc.) confining a water bearing stratum (aquifer). see **Aquifer**

AQUIFER
A water bearing stratum. These can be unconfined (free surface) or confined. A free surface aquifer has an upper limit, the unconfined free surface of the water table. A confined aquifer is overlaid by an impervious aquiclude. see **Aquiclude**.

AREA OF OUTSTANDING NATURAL BEAUTY (AONB) (UK)
Under provisions of the National Parks and Access to the Countryside Act of 1949, certain areas of natural beauty were selected for special protection by local planning authorities.

AROMATIC COMPOUNDS
Organic compounds which have a ring structure. Simplest is benzene, which has the formula C_6H_6. An important derivative is phenol, C_6H_5OH.

ARRHENIUS RELATION
The rate constant (velocity) of a reaction is a function of the heat of activation, the universal gas constant and the absolute temperature. see Q_{10}. see **Van't Hoff rule**

ARSENIC

A brittle poisonous element widely distributed in Nature. It is often found with copper, lead, nickel, iron, cobalt, gold and silver. It occurs most often as sulfide, arsenide, arsenite or arsenate.

ARTICLE 60 (UN CONVENTION ON THE LAW OF THE SEA, 1982)

This section states that installations such as oil drilling platforms which are unused and/or abandoned should be removed in order to insure safety in navigation and that there must be due regard given to fishing and environmental protection. General removal requirements, guidelines and standards were adopted.

ASBESTOS

Generic term for silicate minerals which occur in fibrous form. Chrysotile, due to the length and flexibility of its fibers, can be spun and woven. Other types of asbestos are amphibole-amosite crocidolite, tremolite, autophyllite and actinoite.

ASBESTOS HAZARD EMERGENCY RESPONSE ACT (1986) (US)

Rules and regulations concerning identification, evaluation and control of asbestos containing materials in schools are contained in this Act. The subject of certification of workers is also covered. see **Rip and skip**

ASBESTOSIS

A pneumoconiosis due to inhalation of asbestos dust. Pulmonary fibrosis as a result of inhalation of the chrysotile form of asbestos. Fibers up 200 microns in length and 5 microns in diameter can penetrate the respiratory tract. Inhaled fibers can accumulate in alveolar ducts but cannot penetrate the alveoli.

ASCARIASIS

see **Roundworm**

ASKAREL

see **Polychlorinated biphenyls**

ATMOSPHERIC COMPOSITION

Near the surface of the Earth, the atmosphere is made up of the following (in ppm):

N_2	780 800
O_2	209 400
H_2O	< 35 000
Ar	9300
CO_2	330
Ne	18
He	5
CH_4	2
Kr	1
H_2	0.5
N_2O	0.3
CO	< 0.2
O_3	0.03

ATMOSPHERIC DEPOSITION
see **Acid deposition**. see **Total deposition**

ATMOSPHERIC ENVIRONMENTAL PROBLEMS
Among the pressing problems of the Earth's atmosphere are photo-chemical oxidants, acid deposition, global ozone depletion, toxic chemicals sources and fates, urban and regional haze and particulates.

ATMOSPHERIC INVERSION
When air becomes warmer with distance from the earth's surface, a stable condition exists and there is little or no vertical mixing. The most common case is when air with a weak adiabatic lapse rate is overlain by an inverted layer. Pollutants are then trapped close to the surface. The inversion acts as a lid. Winds are generally weak under this condition. see **Adiabatic lapse rate**

ATMOSPHERIC PRESSURE GRADIENT
Air is compressible, so density, and therefore pressure, decreases exponentially with height above the earth. If air were of constant density, the atmosphere would be about 26 000 ft (7924 m) thick.

ATMOSPHERIC TURBULENCE
Horizontal and vertical eddies which mix contaminated air with surrounding clean air.

ATOMIC ABSORPTION SPECTROPHOTOMETER (AAS)

An instrument for determining the presence and concentration of an element. The system consists of a flame unit, a prism for dispersion and isolation of emission lines and a detector.

ATOMIC ENERGY ACT (US)

A license must be obtained from the Nuclear Regulatory Commission for production or distribution of nuclear materials. A separate license must be obtained for operation of a nuclear power plant. The original Act (1954) placed production and control of nuclear materials under the Atomic Energy Commission.

ATP

see **Adenosine triphosphate**

AUFBAHN

Assimilation.

AUTOECOLOGY

The study of single individuals in relation to ecological processes. see **Ecology**. see **Synecology**

AUTOTROPHIC

Self nourishing. That component of an ecosystem in which light energy fixation, use of simple inorganic substances and buildup of complex substances predominate. see **Photosynthesis**

AVOGADRO'S NUMBER

The number of atoms or molecules present in one gram-molecular weight (6.02×10^{23} atoms or molecules).

B

BACK-MIX REACTOR
see **Completely mixed reactor**

BACTERIA.
Single cell microorganisms which take in soluble food for conversion to new cells. Bacteria exist as single cells, in chains and in clusters. Reproduction is by binary fission. A few have photosynthetic pigment. Bacteria have rod, spherical or spiral shapes.

BACTERIAL DYSENTERY.
see **Shigellosis**

BACTERIOPHAGE
see **Virus**

BAGASSE
Sugar cane stalks. These can be used in biogas production.

BAGHOUSE.
see **Fabric filtration**

BALLAST WATER
This is taken aboard a vessel, for at least part of a voyage, in order to obtain sufficient draft and trim for proper steering and propeller immersion. If ballast tank capacity is inadequate, it is the practice to

add water to empty cargo or fuel tanks. This can result in overboard discharge of contaminated water and transfer of nuisance organisms. Such discharge is now largely outlawed by international agreement.

BANCROFT'S LAW

A surfactant that is principally water soluble disperses oil in water and establishes water as the continuous phase while a surfactant that is principally oil soluble disperses water in oil and establishes oil as the continuous phase.

BAP

see **Benzo-(a)-pyrene**

BARIUM

An active divalent metal. Inhalation of barium can cause the non-malignant lung disease baritosis, which is characterized by fibrous hardening.

BASE FLOW

Normal volumetric flow, derived from soil and groundwater storage, carried in a river, stream or channel.

BAT (BACT)

see **Best available technology**

BATCH PROCESS

Reactants are brought together in a reactor for a suitable time and the products are discharged together at the end of this period. This is contrasted to a continuous process. see **Continuous process**

BATNEEC

see **Best available techniques not entailing excessive cost**

BDAT

see **Best available technology**

BECQUEREL (Bq)

Activity of a radionuclide having one spontaneous nuclear disintegration per second. see **Curie**

BEER'S LAW
The intensity of light incident on an absorbing medium decreases exponentially with increasing concentration of the absorbing medium. Spectrophotometric analysis is based on this law and Lambert's Law. Usually, absorption of light is said to follow the Lambert–Beer Law. see **Lambert's Law**. see **Optical density**. see **Spectrophotometer**

BENTHAL DEPOSITS
Biodegradable material which settles from flowing water forms bottom (Gr., benthos) deposits which undergo microbial action and exert an oxygen demand.

BENZENE
C_6H_6, the parent aromatic molecule, this has three double bonds. Some chlorinated benzene compounds have environmental impact. Polychlorinated biphenyls (PCBs) are particularly important. see **Polychlorinated biphenyls**

BENZO-(A)-PYRENE (BAP)
A polynuclear aromatic hydrocarbon found in cigarette smoke.

BERTALANFFY'S HYPOTHESIS
Microbial growth is the competition between two opposing processes. Aufban is assimilation and Abban is endogenous respiration. The rate of assimilation is proportional to cell protoplasm mass and cell surface area while endogenous respiration is dependent on environmental factors.

BEST AVAILABLE CONTROL TECHNOLOGY
see **Best available technology**

BEST AVAILABLE CURRENT TECHNOLOGY
see **Best available technology**

BEST AVAILABLE TECHNIQUES (TECHNOLOGY) NOT ENTAILING EXCESSIVE COST (BATNEEC) (UK)
Required under the approach of integrated pollution control, this allows control of discharges of material to the air, water and land using currently available technology.

BEST AVAILABLE TECHNOLOGY (BAT)

Techniques for waste treatment which have been shown through actual use to be superior to comparable technologies that are currently considered practical. Also called best available current technology, best available control technology and best demonstrated available technology.

BEST DEMONSTRATED AVAILABLE TECHNOLOGY

see **Best available technology**

BETA RAYS

These are negatively charged particles (electrons) which move at a high velocity, approaching the speed of light. These particles are not as easily attenuated as are alpha particles and have less ionizing power.

B-HORIZON

see **Soil**

BIOACCUMULATION

Increase of concentrations of slowly oxidized or excreted substances in organisms.

BIOAUGMENTATION

Addition of bacterial formulations to wastewater treatment plants which are not performing adequately due to bulking, poor BOD removal due to lipids, lack of nitrification and poor removal of certain compounds, such as phenols.

BIOCHEMICAL OXYGEN DEMAND (BOD)

This is a fundamental measure of the strength of a liquid waste. It is the amount of dissolved oxygen per unit volume (mg/L) necessary to satisfy the metabolic requirements of microorganisms which utilize the waste as food. It is also a measure of the loading placed on the oxygen resources of the receiving water. Unless stated otherwise, reported values are 5-day 20°C (BOD_5). The BOD reaction is idealized as a first order reaction. However, measurement of BOD exertion over an extended period shows that there are two distinct phases. The first, carbonaceous phase, covers about the first ten days. From this point, at which the more easily metabolized food is almost exhausted, the food source is nitrogenous in nature.

BIOCHEMISTRY
The science which deals with control of the actions of living organisms for achievement of desired ends.

BIOCOENOSIS
A community of organisms.

BIODIVERSITY
see **Convention on Biodiversity**

BIOFUELS
Biomass may be transformed by chemical and biological processes to produce energy sources such as methane, ethanol and charcoal. Bio-diesel is the term for fuel for diesel engines.

BIOGAS
Gas obtained from anaerobic digestion of organic wastes and animal excreta. The gas is a source of reliable, inexpensive and convenient fuel for cooking, heating and lighting in Developing Nations. The gas contains 50–60% methane. Slurry waste from the digester is a nitrogen rich fertilizer. The three processes developed for production of combustible gas from biomass or organic waste are hydrogasification, pyrolysis and biogasification.

BIOGASIFICATION
Anaerobic digestion of wastes for production of biogas. Operating conditions are in the range of 30–37°C and little operational skill is required. This is an ideal process for combustible gas production in rural areas. The residual sludge is a nitrogen rich manure suitable for use as a fertilizer.

BIOGEOCHEMICAL CYCLES
Cycles in which the nutrients are transferred from living organisms to the environment and back to living organisms through decomposition processes in or on the soil.

BIOLOGICAL HALF LIFE
The time required for elimination of one half of an ingested substance. Mercury, for example, has a biological life of about 70 days in humans and excretion is primarily through feces. The biological

half life of mercury in large fish can be as much as one to two years. Factors which affect the elimination rate are age, size and metabolism.

BIOLOGICAL MAGNIFICATION
Concentration of substances with successive trophic steps in the food chain. Examples are DDT and radionuclides.

BIOLOGICAL OXYGEN DEMAND
This term has been used by some writers to describe biochemical oxygen demand. see **Biochemical oxygen demand**

BIOLOGICAL POLYMERS
The four major classes are proteins, polysaccharides (carbohydrates), nucleic acids and lipids.

BIOLOGICAL PRODUCTIVITY
Increase in organic material per unit area or unit volume with time.

BIOLOGICAL SOLIDS RETENTION TIME
see **Sludge age**

BIOLOGICAL TRANSLOCATION
A distribution or redistribution of a toxic substance to a specific location in an organism.

BIOLUMINESCENCE
Emission of visible light by living organisms.

BIOMASS
Weight of tissue per unit area or unit volume.

BIOMETHYLATION
see **Methylmercury**

BIOMOVS
see **Biospheric Model Validation Study**

BIOPHAGE
Organisms which obtain their energy from living matter.

BIOREFINERY

An integrated complex which makes a number of products from a variety of feedstocks. These plants make many useful chemicals, including biofuels, from biomass. Biomass consists of grains, husks and other agricultural residues. Biorefineries cannot rely on a single feedstock because of possible unforeseen prices increases.

BIOREMEDIATION

Stimulation of the naturally occurring bacteria to degrade contaminants in soils and groundwater. Microorganisms present in the environment can degrade a great number of organic contaminants and offer the advantage of in-situ treatment rather than moving large quantities of material.

BIOSORPTION

This modification of the activated sludge process is applied to treatment of wastewaters in which much of the BOD is present in suspended or colloidal form. see **Activated sludge**

BIOSPHERE

Soils, seas, seabeds, freshwater bodies, the atmosphere and organisms within them and on them.

BIOSPHERIC MODEL VALIDATION STUDY

An international cooperative effort to test models for prediction of environmental transfer and bioaccumulation of radionuclides and other trace substances. see **Bioaccumulation**

BIOTRANSFORMATION

Metabolism of a substance by an organism.

BLACK LIST (EC)

Substances in this list are considered dangerous due to toxicity and persistence. They are organohalogen compounds, organophosphorous compounds, organotin compounds, carcinogens, mercury and its compounds, cadmium and its compounds, persistent mineral oils and petroleum hydrocarbons, persistent synthetic substances (solid matter) which may sink or float and interfere with use of the water. see **Grey List**. see **Red List**

BLACK LUNG

see **Pneumoconiosis**

BLACK WATER
Wastewater from water closets and latrines.

BOD
see **Biochemical oxygen demand**

BOG
A wetland for which the primary water source is precipitation. These are usually acidic, rich in plant residue and accumulate significant peat deposits.

BORED-HOLE LATRINE
see **Human waste**

BORON
A non-metallic element which occurs only in combination with other elements. The most toxic boron compounds are pentaborane, decaborane and diborane.

BOTTLE BILLS (US)
Legislative efforts at the state level to encourage container recycling and reuse. Deposit and refund procedures are set up. Legislation of this type was strongly opposed by the container industry. Such an approach is now finding favor in some European countries.

BOX AND CAN PRIVY
see **Human waste**

BRACKISH WATER
Water in which the total dissolved solids concentration is between 3000 mg/L and 20 000 mg/L.

BREAK BULK VESSEL
General cargo vessels carrying dried products in raw material as well as finished and packaged form.

BROAD STREET WELL EPIDEMIC
In London in 1854 Dr John Snow demonstrated that an Asiatic cholera outbreak was due to direct flow of sewage contamination into the Broad Street well, the local water supply. He removed the pump handle, effectively dealing with the source of the infection.

BROWNFIELDS

Most abandoned, idled or underutilized properties where expansion or redevelopment is complicated by the potential or confirmed existence of chemicals of concern (ASTM). This is closely the same as the US EPA definition.

BROWNIAN MOTION

A very small particle has a random motion which is caused by incessant bombardment of molecules against the particle.

BUBBLE CONCEPT

see **Emissions trading**

BUFFERS

Substances which resist solution pH changes. The carbonate-bicarbonate system is important in maintenance of relatively constant pH values of natural waters.

BULKING

Failure of activated sludge to flocculate and settle properly.

BULK SAMPLING

Collection of a mixture of wet and dry deposition. There is debate as to the fraction of dry deposition retained by open containers. Variables are wind direction, container and chemical species. This term is also applied to producing a representative sample of non-homogeneous material such as coal and asbestos.

BYPASS

A means of diverting a portion or all of the flow into a wastewater treatment plant during periods of high runoff resulting from a storm. The practice now is to retain the volume bypassed in a detention basin. see **Detention basin**

C

see **Clean Air Act**

CADMIUM

A relatively rare metal not found free in Nature. It is obtained primarily from zinc containing ores and is used for electroplating iron and steel. Pollution control procedures are those applicable to fumes, mists and dusts from the metal refining industry. Inhalation of cadmium causes a chemical pneumonitis and ingestion of cadmium gives severe gastrointestinal irritation.

CALIFORNIA DESERT PROTECTION ACT (US)

Passed at the last minute in 1994, this measure extends federal protection to six million acres of desert in Southern California. It allows continued hunting, mining, grazing and off road vehicle use in areas where these activities are now carried on. The largest land conservation measure since the Alaska National Interest Land Conservation Act, it was held up by a filibuster for political gain. The Act was saved by bipartisan cooperation.

CAP

A cover of clay or other impermeable material over a closed landfill.

CAP AND TRADE

Greenhouse gas producers who exceed certain tonnages of harmful emissions can buy credits from producers that have lowered emissions.

This is practiced in Europe but has been opposed by some business interests and the national government in the United States. see **Emissions trading**

CAPTURE VELOCITY
The velocity at which the energy of motion of an airborne contaminant can be overcome and the contaminant can be drawn into a hood for expulsion from the workplace.

CARBOHYDRATES
Organic compounds of carbon, hydrogen and oxygen in which the hydrogen and oxygen atoms are in the same ratio as in water (2:1). These are classified according to complexity of the structure and are monosaccharides (simple sugars), disaccharides (complex sugars) and polysaccharides. Carbohydrate names end in —ose. Glucose is a hexose ($C_6H_{12}O_6$). Among the disaccharides are sucrose, lactose and maltose, all of which have the formula $C_{12}H_{22}O_{11}$. Among the polysaccharides are starch and cellulose.

CARBONACEOUS OXYGEN DEMAND
The first phase in exertion of Biochemical Oxygen Demand, where the more readily available organic matter (food) is metabolized.

CARBONATE-BICARBONATE SYSTEM
The principal source of alkalinity in natural waters. Dominance of the form is a function of pH. see **Alkalinity**

CARBON CREDITS
see **Kyoto Protocol**

CARBON DIOXIDE
CO_2 is the product of complete combustion of carbon. It is produced naturally by decomposition oxidation of organic matter. Combustion of coal, oil and natural gas produces 90% of anthropogenic CO_2 emissions. CO_2 can heat up the earth's surface by the greenhouse effect. The greenhouse effect is caused primarily by water vapor and carbon dioxide, both strong absorbers of infrared radiation. When absorbed by CO_2 and water vapor, part of the radiation is reflected back to the earth, causing a temperature rise. see **Greenhouse effect**

CARBON 14

A natural isotope used as a tracer in studies in organisms and ecological systems. This isotope also finds application in dating materials and has other uses such as dating of lake sediments.

CARBON FOOTPRINT

The range of carbon emissions, both direct and indirect.

CARBON MONOXIDE

A colorless, odorless gas resulting from incomplete combustion of carbon and carbon compounds. CO production from space heating has been a problem from the dawn of history. Rapid growth of use of the internal combustion engine, particularly in crowded areas with canyon-like streets, has intensified the CO problem. A chemical asphyxiant, CO attaches itself more easily to hemoglobin than does oxygen, forming carboxyhemoglobin (COHb). Cigarette smoke contains 15 000–55 000 ppm of CO and binds to 3 to 10% of the smoker's hemoglobin.

CARBONYL GROUP

A reactive group, $>C=O$, this is fundamental to the chemistry of aldehydes and ketones.

CARBOXYHEMOGLOBIN (COHb).

Carbon monoxide combines readily with hemoglobin and reduces its ability to carry oxygen. Blood hemoglobin has a greater affinity for carbon monoxide than for oxygen (20:1). Formation of carboxyhemoglobin is a reversible process.

CARBOXYL GROUP

This group is present in all organic acids and is usually written as – COOH. Saturated and unsaturated monocarboxylic acids are commonly known as fatty acids. Among the saturated acids, acetic acid is significant in sludge digestion and butyric and valeric acids are associated with quite dreadful odors.

CARCINOGENIC

Cancer causing.

CARDINAL TEMPERATURES

Minimum, maximum and optimum temperatures for functions in an organism.

CASK
A container, usually lead, in which radioactive material is transported.

CASUALTY
Damage to or loss of a ship due to mishaps such as collision or grounding.

CATALYST
A substance which can change the rate of a chemical reaction and can be recovered in the original form at the end of the reaction.

CATALYTIC CONVERTER
Internal combustion engine exhaust gases are passed through a canister packed with catalyst particles for cleaning. Results of such treatment have been acceptable but it has been difficult to maintain long term stability of the units. The need for simultaneous removal of several pollutants has led to interest in dual stage converters. NO_x removal requires reduction while CO requires an oxidation catalyst.

CATCH BASIN
A chamber built at the curbside for admission of surface runoff to a sewer or drain. A sedimentation sump is provided to collect grit and detritus.

CATEGORICAL STANDARDS
These are developed for pollutants that interfere with the operation of a Publicly Owned Treatment Works (POTW), pass through a plant or contaminate the sludge or other residues. Also, substances considered for categorical standards are those for which there is evidence of carcinogenicity, mutagenicity or teratogenicity. see **Sewer ordinance**

CERCLA
see **Comprehensive Environmental Response, Compensation and Liability Act**

CESIUM-137
A major fission product found in fallout and radioactive waste, this radionuclide acts much like potassium. It enters plants through leaves and cattle when they graze the leaves. It appears in cattle muscle tissues and milk. In the human body it is distributed uniformly in muscles. It remains only four months in the human body. Since cesium-137 has a

radioactive half life of 28 years, there is not a great amount of radio-active decay while this substance is in the body.

CESSPOOL
Sometimes called a leaching pit, this is simply a lined or unlined pit into which wastewater is allowed to flow. Solids settle to the pit bottom and water seeps into the ground.

CESTODES
see **Helminths**

CFC
see **Chlorofluorocarbons**

CHELATION
Formation of a cyclic compound in which a metal ion is incorporated. Examples are addition of a chelating (complexing) agent in water softening to mask chemically the hardness ion and detoxification of lead in the human body.

CHEMICAL OXYGEN DEMAND (COD)
A measure of the non-microbial requirement of a waste for oxygen. COD data are important in assessing the strength of an industrial waste where the waste may be toxic or resistant to biological oxidation.

CHEMICAL PRECIPITATION
Addition of divalent or trivalent ions to aid settling of colloidal particles.

CHEMICAL TOILET
see **Human waste**

CHEMOSTAT
A device for controlling a microbial culture at a steady state.

CHEMURGY
The branch of chemistry dealing with industrial utilization of organic raw materials from farm products and the use of renewable resources for materials and energy.

CHERNOBYL INCIDENT
A reactor at the Chernobyl Nuclear Power Station in the former Soviet Union exploded and caught fire in April 1986. It was estimated

that 30 million curies (Ci) (1.1×10^{18} Bq) were emitted. Although about half of the fallout was within 30 kilometers of the accident, the balance was spread over much of Europe. Various estimates of the damage dealt with agriculture and increased human deaths due to cancer.

CHICK'S LAW
The rate of bacterial kill (or die away) per unit time is proportional to the number of organisms yet remaining. This is a first order reaction but departures are common due to variations within the species making up the microbial population. see **First order reaction**

CHLORAMINE
When ammonia is present in a water to which chlorine is added, the ammonia is converted successively to chloramide (monochloramine), chlorimide (dichloramine) and NCl_3 (trichloramine).

CHLORDANE
see **Organochlorine insecticides**

CHLORINATED BENZENE COMPOUNDS
These compounds are used in pesticides and industrial solvents. These are fairly volatile and moderately water soluble. They can be found in wastewaters and can get into groundwater. An example is penta-chlorophenol. PCBs are chlorinated benzene compounds.

CHLORINATION
Application of chlorine to waters to prevent the spread of waterborne diseases. Chlorine reacts with water to form hypochlorous acid, which is the actual substance which kills microorganisms. A chlorine residual is normally maintained in potable water as a defense against extraneous material accidentally introduced.

CHLORINE
A dense greenish-yellow gas with an irritating odor. It was once used as a poison gas in war. Major anthropogenic sources of chlorine include chemical decomposition of chlorofluorocarbons used as refrigerants and propellants. Chlorine is strongly reactive and can deplete the ozone layer. Stratosphere ozone depletion results from photolytic destruction of chlorofluorocarbons (CFCs) and subsequent release of chlorine atoms. $Cl + O_3 = ClO + O_2$. It is estimated that one chlorine

atom has the potential to destroy 105 ozone molecules before reacting with hydrogen to form HCl and being removed from the cycle.

CHLOROFLUOROCARBONS.
Compounds used extensively as aerosol propellants, foam blowing agents and refrigerants. The two most commonly used were $CFCl_3$ (F11, US: R11, UK) and CF_2Cl_2 (F12, US: R12, UK). These are quite stable and eventually reach the stratosphere. Here they are photolyzed by high energy ultraviolet radiation to provide free chlorine atoms which can destroy ozone. CFCs have been outlawed in many industrialized countries. CFCs have been replaced by hydrochlorofluorocarbons (HCFC), which are much less damaging to the ozone layer.

CHLOROPHYLL
Green and yellow pigments found in green plants. Incident light energy splits electrons from chlorophyll. These are attached to a coenzyme and produce a hydrogen atom. The hydrogen atom is the actual fuel for a plant cell. see **Coenzymes**

CHLOROSIS
Whitening of green plants due to reduction in chlorophyll resulting from pollution, acid rain or disease.

CHOLERA
An enteric infection spread by contaminated water. Prevalent in less developed countries, it is rare in more advanced societies due, to a great degree, to chlorination of water supplies.

C-HORIZON
see **Soil**

CHP
see **Combined heat and power**

CHROMIUM
Widely used in plating and alloys, this hard white metal in the hexavalent form (Cr^{+6}) is toxic to humans and animals and can upset biological wastewater treatment systems. The hexavalent form is banned by sewer ordinances.

CITES
see **Convention on International Trade in Endangered Species**

CITRIC ACID CYCLE
The terminal oxidation scheme for acetate, this is the fundamental process in microbial metabolism. The sulfa drugs were effective against disease causing organisms because they interrupted the citric acid cycle for those organisms which could not synthesize their own para-amino-benzoic acid (PABA).

CLADDING
The non-corrodible covering, usually zirconium or stainless steel, in which the fuel rods of a nuclear power reactor are contained.

CLEAN AIR ACT (CAA) (US)
Federal nationwide ambient air quality standards for conventional air pollutants are provided for in this Act. Provision is made for attainment of these standards by reducing pollution from stationary and mobile sources. The Act was amended significantly in 1990.

CLEAN WATER ACT (CWA) (US)
This law sets a goal of fishable/swimmable water and attempts to achieve this by construction grants for publicly owned treatment works, a permit system for point sources of pollution, requiring the equivalent of secondary treatment and areawide water quality management. Included are provisions for wetlands protection, sludge disposal, ocean discharge and estuary protection. Liability for oil spills is covered under Section 311.

CLOSURE PLAN (US)
Documentation prepared to guide deactivation, stabilization and monitoring of a facility under RCRA.

CLUB OF ROME
A group of interested professionals, of diverse disciplines, set up in 1968 to study the relationships among various components of the human social systems. The ultimate aims are predictions of the results of present policies and formulations of alternative policies.

COAGULATION

In treatment of water, salts of divalent and trivalent ions are added to neutralize charges on colloidal particles. Coagulation, strictly speaking, is neutralization of charge.

COAL GASIFICATION

Coal is heated in the presence of steam and an oxygen containing gas yielding, typically, hydrogen, carbon monoxide and variable amounts of light hydrocarbons.

COASTAL ZONE MANAGEMENT ACT (CZMA) (US)

Assistance is provided to state governments for development of Coastal Zone Management Plans. The primary function of these plans is to assist orderly and environmentally sound development of these sensitive areas.

COD

see **Chemical oxygen demand**

CODEX ALIMENTARIUS

A commission of the Food and Agricultural Organization (FAO) which has published principles concerning use of food additives.

COEFFICIENT OF VARIATION

A statistical measure of scatter of data about the mean. The standard deviation divided by the mean.

COENZYMES

Complex heat stable compounds which must be present before certain other enzymes can act. These can function as carriers of hydrogen and can be donors or acceptors of hydrogen.

COGENERATION

see **Combined heat and power**

COLIFORMS

see **E. coli**

COLLOID

A very small particle dispersed in a medium. Colloidal particles are usually in the range of 1 micron to 100 microns. These particles have large surface area in relation to mass and this property is of great significance. Charges collect on the surfaces and colloidal dispersions may be quite stable. Colloidal particles formed during water treatment processes are removed by addition of electrolytes which supply ions of opposite charge. Coagulation is neutralization of charge. Removal of color from natural waters requires neutralization of particle charge. An emulsion is formed by dispersion of one liquid in another liquid. Of greatest interest is dispersion of oil in water.

COLOR
The presence in water of tannins, humic acid and humates resulting from decomposition of lignins. These are considered to be the principal sources of colors in natural water. Natural color exists mostly as negatively charged colloidal particles. Objection to color is ordinarily on esthetic grounds. The standard color unit is 1 mg/L of platinum, expressed as potassium chloroplatinate (K_2PtCl_6).

COMBINED BULK CARRIER
A vessel which can carry dry and liquid cargos within the same cargo hold spaces.

COMBINED HEAT AND POWER (CHP)
Use of the same generating plant for simultaneous production of electricity, process heat and/or steam.

COMBINED SEWER
One which carries both wastewater and stormwater. Design is on the basis of the storm flow.

COMETABOLISM
Also called co-oxidation, this is a mechanism by which microorganisms can chemically alter a compound without deriving useful carbon or energy for growth from the compound. This seems to be an important route in degradation of recalcitrant compounds, such as pesticides.

COMPACTION
The most commonly used means of volume reduction for solid wastes. High pressure compaction (up to 3000 psi) can produce a closely inert fill material or building product.

COMPLETELY MIXED REACTOR

Contents of the reactor are of constant composition throughout. Another name is Back-Mix Reactor.

COMPLEX

see **Chelation**

COMPLIANCE AGREEMENTS (US)

Agreements between regulators and regulated groups which set standards and schedules for compliance with environmental laws. These include consent orders and compliance agreements, federal facilities agreements and federal compliance agreements.

COMPOSITE SAMPLES

These are taken at a constant time interval and then mixed in proportion to the flow rate at the time of sampling to give a representative sample for the whole time period. Also called an integrated sample.

COMPOSTING

Decomposition of semi-dry organic material by microorganisms. The process can be carried out in both batch and semicontinuous operations. It is necessary to turn the windrows (stacks) periodically for mixing and aeration. A typical process shows a temperature rise above ambient and it takes several months to achieve a stable humus-like final product. Volume reduction is in the range of 30–50%.

COMPREHENSIVE ENVIRONMENTAL RESPONSE, COMPENSATION AND LIABILITY ACT (CERCLA) (US)

This 1981 Act, commonly called Superfund, established a fund to finance cleanup activities at currently operating and abandoned facilities. EPA can use fund monies to clean up a contaminated site and then proceed against responsible parties. Current owners, owners or operators of the facility at the time of hazardous substance release and generators and transporters of the substances can all be held liable. Innocent purchasers who made a reasonable investigation are insulated from action. A National Priority List was specified. The 1986 amendments deal with the Superfund sites to be cleaned up. The EPA is required to select remedial actions for permanent solutions and use alternative technologies for resource recovery. Preference is given to treatment that permanently reduces volume, toxicity or mobility of toxic substances. Offsite transport and disposal without treatment is the least desirable alternative. On-site remedial actions must conform to all relevant requirements under the Safe Drinking Water Act and the Clean Water Act. Title III of the Amendments sets forth a regulatory program that requires disclosure to workers and the general public about dangers of hazardous chemicals and mandates development of emergency response plans for chemical emergencies. Included is a section which allows designated state and federal agencies to seek damages for loss of natural resources as a result of release of hazardous substances.

CONCENTRATION
The quantity of a substance present per unit volume, expressed as mol/L.

CONFERENCE ON ENVIRONMENT AND DEVELOPMENT (UN)
Called the Earth Summit, the basic message of this 1992 Conference was that necessary changes in attitude and behavior must be made to bring about needed changes in use of the planet. Stressed were poverty and stress on the environment resulting from excessive consumption by more affluent countries.

CONGENERS
Different molecular combinations. Congeners which comprise the series of compounds with the same number of a given atom are an homologous series.

CONSECUTIVE REACTION
The products of one reaction become the reactants in a following reaction. Examples are the oxygen deficit caused by pollution in a stream and bacterial nitrification of ammonia.

CONSTRUCTED WETLANDS
These do not have constraints on discharge to a natural wetland. Natural wetlands are considered receiving waters. The two types of constructed wetlands are free water surface (FWS) and subsurface flow (SF). In the FWS system the flow is horizontal and continuous. Pretreated water flows slowly through roots and stems of emergent vegetation. The SF system has a channel filled with porous media (rocks, gravel, etc.) and a barrier to prevent seepage. The media support the root system of emergent vegetation. Flow is horizontal. In both systems the plants extract nutrients from the wastewater.

CONSUMER PRODUCT SAFETY ACT (US)

The Consumer Product Safety Commission, created under this Act, is authorized to impose performance standards for packaging and labeling requirements on consumer products in order to prevent unreasonable risk of injury. A cost/benefit analysis is required before the Commission can place a restriction on a product. A hazardous product can, however, be banned from commerce.

CONSUMER PRODUCT SAFETY COMMISSION

see **Consumer Product Safety Act**

CONSUMERS

Heterotrophic organisms which must depend for food energy on organic material already synthesized in living or dead organisms. see **Heterotrophic**

CONTACT STABILIZATION

see **Biosorption**

CONTINUITY

Mass entering a system per unit time is equal to the mass leaving per unit time plus the mass stored per unit time.

CONTINUOUS PROCESS

One in which reactants enter the reaction chamber and, at the same time products are being removed.

CONTROL OF POLLUTION ACT (COPA) (UK)

General parts of this 1974 Act are 1 – waste disposal, 2 – water pollution, 3 – noise, 4 – air pollution and 5 – miscellaneous.

CONVENTION FOR PREVENTION OF MARINE POLLUTION BY DUMPING FROM SHIPS AND AIRCRAFT (OSLO CONVENTION)

see **Convention for Protection of the Marine Environment of the North-East Atlantic**

CONVENTION FOR PREVENTION OF MARINE POLLUTION FROM LAND-BASED SOURCES (PARIS CONVENTION)

see **Convention for Protection of the Marine Environment of the North-East Atlantic**

CONVENTION FOR PROTECTION OF THE MARINE ENVIRONMENT OF THE NORTH-EAST ATLANTIC (OSPAR)

This agreement replaces the Oslo Convention and the Paris Convention. The intent is to provide a comprehensive approach to all pollution sources which can affect the maritime area and other matters relating to the maritime environment.

CONVENTION ON BIODIVERSITY (UNEP)

An international treaty on preservation of species diversity signed by many countries in 1992. Funds were provided for conservation and research projects in Developing Countries. Provision was made for payment of royalties for genetic material culled from these lands. An on-line electronic data base on plants and animals is being established.

CONVENTION ON INTERNATIONAL TRADE IN ENDANGERED SPECIES (CITES)

A treaty signed by more than 80 countries, which prohibits international trade in certain rare animals and plants.

CONVENTION ON WETLANDS

This was adopted in the Iranian city of Ramsar in 1971 and came into force in 1975. It was the only global treaty dealing with a single eco-system. The Convention addressed swamps, marshes, lakes, rivers, wet grasslands, peatlands, oases, estuaries, deltas, tidal flats, mangroves, coral reefs and human made sites such as ponds, rice paddies and reservoirs. It established criteria for inclusion in the List of Wetlands of International Importance (Ramsar List).

COOLING TOWERS

Hot water discharged to a stream, shallow lake or coastal water may have an adverse effect. Instead, water is cooled by evaporation in specially designed towers. This water may later condense to form atmospheric vapor plumes. Cooling towers for air conditioners have been found to contain organisms such as *Legionella pneumophilia*.

COPA

see **Control of Pollution Act**

COPPER

An excellent conductor of electricity and heat, this metal is toxic to fish and other animals at elevated concentrations in the divalent form (Cu^{+2}).

Copper sulfate ($CuSO_4$) has been used in control of algal blooms. Sources of copper pollution include mining, ore processing or plating.

CORIOLIS ACCELERATION

This is often called a force but is actually an acceleration due to rotation of the earth. It causes a motion deflection to the right in the northern hemisphere and to the left in the southern hemisphere. This has a fundamental effect on movement of atmospheric air masses.

CORRELATION COEFFICIENT

A measure of correlation among variables, the most commonly treated is the case of two variables. The square of the correlation coefficient gives the proportion of the variance associated with one of the variables which can be predicted from knowledge of the other variable.

COST/BENEFIT

(1) (US). A method used by the US Army Corps of Engineers to evaluate the worth of a proposed project. The estimated cost is compared to the value of the expected benefits. (2) (US). An imprecise evaluation of the estimated costs of environmental laws and regulations. This is regarded as a political effort to circumvent existing environmental efforts. (3) (US). Analysis required of the Consumer Product Safety Commission before restrictions can be placed on a product.

COUNTER ION

These divalent and trivalent ions are used as coagulants due to their ability to neutralize colloidal charges.

CRITERIA POLLUTANTS

Those which present a threat to public and welfare. see **Hazardous pollutants**

CRITICAL DEFICIT

In a watercourse, when the concentration of dissolved oxygen is at a minimum, the deficit, the difference between the saturation value and the actual D. O. concentration, will be greatest. see **Deficit**. see **Oxygen sag equation**

CRITICAL FLOW

An hydraulic phenomenon which is the basis of flow measurements. see **Parshall flume**

CROWN CORROSION

Hydrogen sulfide produced in anaerobic decomposition of sewage is taken into solution in the moisture on the walls of a sewer. The moisture can remain on the inner top (crown) of the pipe. The sulfur oxidizing *Thiobacillus* turn the hydrogen sulfide into sulfuric acid. In concrete sewers the sulfuric acid reacts with lime in the sewer to form calcium sulfate. Structural strength of the pipe is reduced.

C.t VALUE

A measure of the adequacy of disinfection, the product of C, disinfectant concentration in mg/L, and contact time *t*, in minutes.

CURIE (STANDARD) (Ci)

3.7×10^{10} disintegrations per second of a radioactive substance. More commonly used are the millicurie, microcurie and picocurie. The standard now is the Becquerel (Bq). 1 Ci = 3.7×10^{10} Bq. see **Becquerel**

CYANIDE (CN⁻)

An extremely poisonous substance, this can appear in industrial effluents. It is prohibited from discharge to municipal collection systems.

CYANOPHYTA

see **Algae**

CYCLE

see **Biogeochemical cycle**. see **Citric acid cycle**. see **Federal driving cycle**. see **Hydrologic cycle**

CYCLONE

One of the simpler and cheaper dust collectors, this equipment has low efficiency except when used with coarse dust. Gas with entrained dust enters tangentially at 50–100 ft/sec and flows downward in a helical path in an annulus, then upward in the center to an outlet. Dust particles are driven to the walls by centrifugal force of the order of 100 times the force of gravity (100 g) and exit from the dust outlet at the bottom of the cyclone. These units can be installed in series, either with other cyclones or as precleaners before other removal devices.

D

DAF
see **Dissolved Air Flotation**

DALTON'S LAW (OF PARTIAL PRESSURES)
In a mixture of gases, each gas exerts pressure independent of the others. The partial pressure of each gas is proportional to the percentage of the total volume occupied by that gas.

DARCY'S LAW
A description of the relation in a granular material among flow velocity, permeability and hydraulic gradient. This is applicable only below a Reynolds number of 0.1. see **Reynolds number**

DARCY-WEISBACH EQUATION
An expression for calculation of energy loss in fluid flow in terms of velocity, pipe geometry and length. The friction factor, f, is a function of pipe roughness and Reynolds number. The Moody diagram is used for determination of f by iteration. see **Moody diagram**

dB
see **Decibel**

DDE
A persistent metabolic product resulting from action of soil organisms on DDT.

DDT
see **Dichlorodiphenyltrichloroethane**

DEADWEIGHT TON
The displacement measure of the weight of a vessel.

DECHLORINATION
When large amounts of chlorine are added to water for taste and odor control, unwanted residuals can be removed by aeration, passage through beds of granular activated carbon or addition of reducing chemicals such as sulfur dioxide, sodium bisulfite or sodium sulfite.

DECIBEL (dB)
Sound pressure level. The reference pressure is 20 micropascals, corresponding to the normal threshold of hearing at about 1000 Hz. This value is 0 dB. The audible range covers about 120 dB and the scale is logarithmic. Weighting schemes have been introduced. The most common is the A-scale, a frequency weighted unit used for noise measurements. This scale, dB(A), corresponds to the frequency response of the ear and correlates well with loudness.

DECOMMISSIONING
The process of removing a nuclear facility from operation. Included can be decontamination, dismantling, entombment or conversion to another use.

DECOMPOSER
Consumers, mostly bacteria and fungi, which break up complex organic substances of dead matter. Some of the decomposition products are taken into the organism's own protoplasm and the remainder becomes available to consumers.

DEFICIT
The difference between the saturation concentration of a substance and the actual concentration. The forces driving restoration of the saturated condition are the deficit itself, concentration gradient and turbulence. see **Critical deficit**. see **Oxygen sag equation**

DEGREE-DAY
A means of assessing fuel requirements and heating efficiency. In the UK, heating degree-days are the product of the difference between the

average temperature and 16°C, times the number of days of record. In the US, total degree-days are the difference between the average temperature and 65°F, times the total number of record days. Cooling degree-days are calculated in a similar manner.

DEIONIZATION
see **Ion exchange**

DELANEY AMENDMENT (US)
This amendment to the Food and Cosmetics Act states that no additive will be deemed safe if it is found, under appropriate test conditions, to induce cancer in animals or humans.

DE MINIMUS LIMIT
A contaminant level below which effects are negligible.

DENATURE
Bring about irreversible change.

DENITRIFICATION
Conversion of nitrate ion in wastewater to nitrogen gas.

DENTAL FLUOROSIS
Mottling of tooth enamel due to prolonged ingestion of water high in fluorides. As a result of extensive study of this phenomenon, it was concluded that it is desirable for public water to have a fluoride concentration of about 1 mg/L in order to control dental caries.

DEOXYRIBONUCLEIC ACID (DNA)
The carrier of genetic information. The DNA molecule is composed of two long stranded chains in a double helix configuration. see **Ribonucleic acid**

DEPARTMENT FOR ENVIRONMENT, FOOD AND RURAL AFFAIRS (DEFRA) (UK)
This combined Department brings together efforts dealing with air, water, food, farming and the rural economy. Its functions are those of several separate Departments. One major objective is to bring about sustainable development.

DEPARTMENT OF THE ENVIRONMENT (DOE) (UK)
see **Department for Environment, Food and Rural Affairs**

DEPOSITION VELOCITY
In removal of material from the atmosphere by dry deposition, this is the ratio of the deposition rate and the concentration.

DESALINATION
Preparation of potable water from brackish or saline water. Water can be separated from salt or salt from water, depending on the process used. The most commonly used methods are distillation, electrodialysis and reverse osmosis. Ion exchange and freezing have not yet been proved economical for significant desalination application. Disposal of brine as a waste product is a problem associated with all desalination processes. One serious economic problem often associated with desalination is the cost of pumping water from the point of production to the point of consumption. see **Electrodialysis**. see **Ion Exchange**. see **Reverse osmosis**

DES
see **Diethylstilbestrol**

DESTRUCTIVE DISTILLATION
see **Pyrolysis**

DETENTION BASIN
A structure into which flood flows are brought in order to lessen the magnitude of the maximum instantaneous discharge downstream. see **Flood routing**

DETENTION TIME
see **Hydraulic retention time**

DETERGENT
A substance used for cleaning. Basic ingredients are surface active agents (surfactants), which are large polar molecules. see **Synthetic detergents**

DETRITUS
A mixture of grit and organic material collected in a grit chamber. It is unsuitable for use as a fill material because of its putrescibility.

DEVELOPING NATIONS

Also called Third World, these are defined as having a low standard of living, low per capita energy consumption, low per capita income, low per capita food consumption, low life expectancy, high birth and death rates and poor educational and health care provisions.

DEW POINT

The temperature at which water vapor starts to condense.

DIATOMACEOUS EARTH FILTER

Microscopic skeletons of diatoms from prehistoric seas used for a filter bed for production of potable water and polishing of wastewater effluents.

DIAUXIE

Two microbial cultures supplied with limiting substrates for both food and energy will, in many cases, utilize one until exhausted and then, after a short lag phase, begin to use the second.

DICHLORODIPHENYLTRICHLOROETHANE (DDT)

A widely applied chlorinated pesticide used during and after World War II. It has been implicated in many environmental problems, among which is the "Thin Eggshell Syndrome."

DIELDRIN

see **Organochlorine insecticides**

DIETHYLSTILBESTEROL (DES)

A growth promoting hormone for cattle, this has been shown to be carcinogenic in laboratory animals.

DIFFERENCE LIMEN

The minimum energy change required to perceive a change in stimulus.

DIFFUSION

Molecular mixing due to random motion of molecules.

DIMETHYL SULFIDE

Dimethyl sulfoxide is converted by microorganisms in wastewater to dimethyl sulfide, the compound responsible for the rotten cabbage odor associated with wastewater treatment plants. see **Dimethyl sulfoxide**

DIMETHYL SULFOXIDE

This common industrial solvent is not itself objectionable but is converted by microbial action to dimethyl sulfide, which is the source of the odor of rotten cabbage around sewage treatment works. see **Dimethyl sulfide**

DINOFLAGELLATES

see **Protozoa**. see **Red Tide**

DIOXIN

There are 75 compounds in this family of chlorinated compounds. Low temperature or incomplete combustion of organic chlorine compounds can produce dioxins. The dioxins of greatest environmental interest are polychlorodibenzo-p-dioxin (PCDD) and 2,3,7,8-tetra-chlorodibenzo-p-dioxin (TCDD). TCDD was present as a contaminant in Agent Orange, a chemical defoliant widely used in Vietnam and to which many people were exposed. see **Agent Orange**

DISINFECTION

Destruction of microorganisms but not bacterial spores.

DISPERSANT

A substance which breaks spilled oil into fine droplets. It was thought that dispersants themselves were more toxic to marine organisms than was the oil itself. However, it has been found that aromatic solvents used to remove tar accumulations are actually the toxic agents.

DISPOSAL BY DILUTION

Treatment of effluent to such a degree that discharge to the environment does not bring about undesirable effects. Effluent standards are, in effect, based on disposal by dilution.

DISSOLVED AIR FLOTATION (DAF)

This process removes suspended solids, oil and grease by means of small air bubbles rising through the wastewater stream.

DISSOLVED LOAD

The part of the total load carried by a river that is in true solution, as opposed to that which is in suspension. The principal ions making up the dissolved load are chloride, sulfate, bicarbonate, sodium and calcium.

Carbonate is not significant since the pH is close to neutral (7). At low flows, when the main recharge source is groundwater, the dissolved load is at a maximum.

DISSOLVED OXYGEN (DO)
The amount of molecular oxygen in true solution. Under normal conditions water can hold 10 mg/L DO at saturation.

DIURNAL
Night and day cycles. Algal production of oxygen during the day and CO_2 release at night is an example.

DL
see **Difference limen**

DMSO
see **Dimethyl sulfoxide**

DNA
see **Deoxyribonucleic acid**

DO
see **Dissolved oxygen**

DOSE-RESPONSE
An increased response to a chemical with increasing concentration of the chemical.

DOUBLE BOTTOM
On some tankers the cargo tank bottom is the bottom of the vessel. If the hull is ruptured, the cargo can flow directly into the sea. A double bottom involves provision of a space between the cargo tank and the skin of the ship.

DOWNWELLING
Winds blowing along the coast with the shore to the right in the northern hemisphere and to the left in the southern hemisphere cause surface currents to move toward the shore. Nearshore surface waters are transported downward and in the offshore direction. see **Upwelling**

DRINKING WATER STANDARDS (EU)

Twenty six parameters are mandated for safe drinking water. They are, in general, comparable to WHO guidelines. The values can change in the light of new research. see **Public Health Service Drinking Water Standards**. see **Safe Drinking Water Act**

DROWNED RIVER VALLEY

A river valley with a relatively wide coastal plain which has been inundated by rising ocean levels. An example is Chesapeake Bay.

DRY BULK CARRIER

Vessels engaged primarily in transport of commodities such as grain, coal and ore.

DRY DEPOSITION

1   All materials deposited from the atmosphere in the absence of precipitation.
2   The process of such deposition.

DUE DILIGENCE

see **Environmental due diligence**

DUPUIT EQUATION

An idealized description of radial flow into a well.

DUST

Airborne particulate matter between 1 and 75 microns in diameter.

DWT

see **Deadweight ton**

DYSTROPHIC

Greatly reduced or disturbed system productivity.

E

EARTH RESOURCES TECHNOLOGY SATELLITE (ERTS)
A series of satellites put into orbit to gather and relay information on land use, forests, water resources and energy use. Renamed LAND-SAT to differentiate from SEASAT, which is only for ocean surveys.

EARTH SUMMIT
see **Conference on Environment and Development**

ECO-AUDIT (EU)
A program within the European Union that permits companies to have environmental management programs verified by outside auditors.

E. (ESCHERICHIA) COLI
A member of the *Enterobacteriaceae* family, this organism is of fecal origin, being a normal inhabitant of the human intestine, and is used as an indicator organism for sewage pollution. In testing for coliforms, the lactose broth test is presumptive but must be followed by a confirmatory test because soil dwelling organisms such as *A. aerogenes* can give a false positive. Brilliant green bile broth (BGB) is used for the confirmatory test. Consumption of meat or vegetables infected with *E. coli* can cause intestinal upsets and, in some cases, fatal results. This usually is due to improper handling of raw food. Irrigation of crops by inadequately treated wastewater has been implicated.

ECOLOGICAL MAGNIFICATION
see **Biological magnification**

ECOLOGICAL SYSTEM
see **Ecosystem**

ECOLOGY
The study of relationships between organisms and the physical and chemical environment and interrelationships among organisms. Odum defines ecology as environmental biology.

ECOSYSTEM
Any unit, including all the organisms, in a given area, interacting with the physical and chemical environment so that energy flow leads to a clearly defined trophic structure. Structural components of an ecosystem include inorganic substances, organic compounds, climate regimes, producers, consumers, phagotrophs and saprotrophs. Processes are energy flow circuits, food chains, diversity patterns, nutrient cycles, development, evolution and control.

ECOTOXICOLOGY
This branch of ecology is concerned with chemical effects on whole ecosystems. Human toxicology focuses on the individual while ecotoxicology deals with communities and populations.

EDTA
see **Ethylenediaminetetraacetate**

EFFECTIVE STACK HEIGHT
The actual height of a stack plus additional effluent rise due to buoyancy, often expressed in term of a velocity head ($v^2/2g$).

EGESTION
Elimination of waste products (feces).

E. (ENTAMOEBA) HISTOYTICA
The causative organism of amoebic dysentery.

EIA
see **Environmental Impact Statement**. see **Finding of no significant impact**

EIS
see **Environmental Impact Statement**

EKMA
see **Empirical Kinetic Modeling Approach**

EKMAN SPIRAL
Water velocity decreases exponentially with depth from a maximum at the sea surface and the surface velocity is displaced 45° to the right of the generating wind in the northern hemisphere and 45° to the left in the southern hemisphere.

EKMAN TRANSPORT
see **Ekman spiral**

ELECTROCOAGULATION
see **Wastewater coagulation**

ELECTRODIALYSIS
Current is impressed across electrodes placed in water. Anions and cations migrate to the electrodes. By placing alternating anionic and cationic membranes, a series of concentrating and diluting compartments is created.

ELECTROMAGNETIC SPECTRUM
This extends over the range of wavelengths from 10^{-12} cm to 10^{10} cm The shortest wavelengths are generated by cosmic and X-rays and the longer wavelengths are associated with power generation and microwaves. Ultraviolet, visible and infrared radiations are in the intermediate range.

ELECTROSTATIC PRECIPITATOR
Stack gases may contain significant amounts of unwanted particulates. These can be removed by passage through a high voltage electric field (12 kV–75 kV). Particles take on a surface charge and are then attracted to collector electrodes which have an opposite charge. It is necessary to remove accumulated dust by mechanical means. Up to 99+% of some particulates can be removed in this manner.

ELEPHANTIASIS
see **Filariasis**

EL NINO

Waters off the coasts of Ecuador and Peru are normally cool due to upwelling in the region. However, when there is alteration of air mass circulation over these areas by interruption of the seasonal shift in atmospheric pressure between the Australian Indian Ocean region and the southeastern Pacific, the normally cooler waters become warm, giving rise to abnormal rainfall variations on a global scale. see **La Nina**

EL NINO/SOUTHERN OSCILLATION (ENSO)

The coupled ocean–atmosphere which includes both El Nino and La Nina.

EMERGING CONTAMINANTS

The US EPA has issued a list of 129 Priority Pollutants. It is now thought that it may be necessary to add to this list. This is due, in part, to advances in analytical instrumentation. It is now possible to detect compounds not previously found.

EMERGING POLLUTANTS

see **Emerging contaminants**

EMPIRICAL KINETIC MODELING APPROACH (EKMA)

A lumped chemical kinetic process model used in generating ozone isopleths for evaluation of control strategies. Calculations are performed using a fixed set of assumptions and varying initial concentrations of reactive organics and nitric oxide. The design value for ozone is the second highest hourly ozone average concentration observed in the location.

EMISSIONS TRADING SCHEMES (EU)

This has been in operation since 2005 and is set by a European Directive. Details are set in National Allocation Plans (NAPs). The EU ETS Directive sets key criteria to which NAPs must adhere and these plans are then assessed. The key considerations for overall emissions caps are national progress toward Kyoto/EU-burden sharing targets, projected emissions development, consistency with other legislation and non-discrimination among companies and sectors.

EMISSIONS TRADING (US)

A plant complex is allowed to decrease pollution from one source and increase pollution from another source, as long as the total result is the same or less than previous results.

EMSCHER TANK
see **Imhoff tank**

EMULSION
Dispersion of one liquid in another.

ENDANGERED SPECIES
A plant or animal species which may not be able to reproduce in sufficient numbers to survive.

ENDANGERED SPECIES ACT (US)
Federal agencies are required to make certain that any actions carried out, funded or authorized by them are not likely to jeopardize continued existence of any listed endangered or threatened species, or result in destruction or adverse changes to its habitat.

ENDEMIC
Persistent presence of a disease in an area.

ENDOGENOUS RESPIRATION (METABOLISM)
Organic food in solution is in equilibrium with the microorganisms and growth has ceased. The microorganisms metabolize their own protoplasm and feed slowly upon the food left in solution. Microbial mass and food concentration remain essentially constant in this phase.

ENDOSPORES
The resting (spore) form of normal vegetative cells.

ENDRIN
see **Organochlorine insecticides**

ENERGY
The ability to do work.

ENERGY FLOW
Efficiency of energy transfer from one trophic level to another.

ENERGY INTENSITY
The energy produced per unit of Domestic Gross Product. The energy declines as a country shifts from an industrial to a service economy or uses its resources more efficiently.

ENVIRONMENT
All influences external to an organism.

ENVIRONMENTAL CONTAMINATION
The presence of substances at concentrations which do not pose any risk or hazard to the environment. At higher concentrations these same substances may threaten health or well being of plants or animals.

ENVIRONMENTAL DEFENSE FUND (EDF) (US)
A non-profit group dedicated to maintenance of a clean environment, the EDF brings suits when environmental issues arise.

ENVIRONMENTAL DUE DILIGENCE (US)
Investigation of the environmental integrity of real estate. This is important due to liability provisions under Superfund.

ENVIRONMENTAL IMPACT ASSESSMENT
see **Environmental Impact Statement**. see **Finding of no significant impact**

ENVIRONMENTAL IMPACT STATEMENT (US)
A report prepared to assess the expected effects, both positive and negative, of a proposed project. While the National Environmental Policy Act (NEPA) restricted the requirement of an environmental impact statement to federal projects, many state and local authorities now require such documents. In general, adverse statements do not, in themselves, bring about acceptance or rejection of the proposed projects. The fundamental intent of these reports should be avoidance of environmental mistakes before they happen or are caused. see **National Environmental Policy Act**

ENVIRONMENTAL JUSTICE
The principle that all people have the right to be protected against environmental pollution and to receive a fair share of environmental benefits.

ENVIRONMENTAL LAPSE RATE
see **Adiabatic lapse rate**

ENVIRONMENTAL PROTECTION ACT (UK)
This is the fundamental environmental legislation in the UK and mandates integrated pollution control. Important parts of the Act include

Sec. 29 which defines environment, Sec. 35 dealing with waste management licenses, Sec. 50 covering waste disposal plans and Sec. 75, wastes.

ENVIRONMENTAL PROTECTION AGENCY (US)
Formed in 1970 by a reorganization plan, powers were transferred to the new agency from the Department of the Interior, Department of Agriculture and Department of Health, Education and Welfare (now Department of Health and Human Services). Functions dealing with environmental matters were assumed by the EPA. It is not represented at Cabinet level.

ENVIRONMENTAL SANITATION
The control of community water supplies, excreta and wastewater disposal, refuse disposal, vectors of disease, housing conditions, food supplies and handling, atmospheric conditions and safety of the working environment (World Health Organization).

ENVIRONMENTAL SITE ASSESSMENT (US)
Investigation of property for contamination resulting from use by previous owners. This is of importance due to liability provisions of CERCLA.

ENVIRONMENTAL TOBACCO SMOKE
A mixture of sidestream smoke from the lighted end of a cigarette and mainstream smoke exhaled by the smoker.

ENZYME
Temperature-sensitive catalysts of organic nature which are produced by living cells and can act within or outside the cell. Enzymes are proteinaceous. Enzymes which catalyze hydrolytic reactions are known as hydrolases and those that catalyze the rupture of linkages are known as demolases or respiratory enzymes. Enzyme names end in -ase.

EPA
see **Environmental Protection Act. see Environmental Protection Agency**

EPIDEMIC
Occurrence of a disease or infection in excess of the expected rate.

EPIDEMIOLOGY

Public health detective work. The field of public health concerned with relationships among various factors and conditions which determine frequency and conditions of an infectious process, disease or physiological state in a human community.

EPILIMNION

The upper circulation zone of a lake or reservoir when the system is stratified.

EQUIVALENCE POINT

Appropriate end points reached in titration.

EQUIVALENT SURFACE AREA DIAMETER

The diameter of a sphere having the same surface area as that of the particle under consideration.

EQUIVALENT VOLUME DIAMETER

The diameter of a sphere having the same volume as the particle under consideration.

ERGONOMICS

The study of the design of the workplace from the standpoint of the worker.

EROSION

Transport of surface particles by weathering, flowing surface water, rain and wind.

ERTS

see **Earth Resources Technology Satellite**

ERYTHEMA

Reddening of the skin due to exposure to radiation.

ESCHERICHIA COLI

see *E. coli*

ESTERS

These compounds are formed by reactions between organic acids and alcohols. They correspond to salts in inorganic chemistry.

ESTUARY

A semi-enclosed body of water having free connection with the open sea and within which the water is measurably diluted with fresh water derived from land drainage.

ETBE

see **Reformulated gasoline**

ETHYLENE ($H_2C=CH_2$)

A colorless gas of the olefin series. Not toxic to humans or animals but is the only gas found to have adverse effects on plants at ambient concentrations of 1 ppm or less. It is the largest volume organic chemical produced today.

ETHYLENEDIAMINETETRAACETATE (EDTA)

A chelating agent which forms stable complex ions with Ca^{+2} and Mg^{+2}, the principal hardness causing ions.

ETS

see **Environmental tobacco smoke**

E. (EBERTHELLA) TYPHOSA

The causative organism of typhoid fever. The disease is spread primarily by intestinal discharges. Prevention of the spread of typhoid was the basis of water sanitation practices for many years. Current thinking now deals with protection of oxygen resources of the receiving waters into which the wastes are discharged.

EUBACTERIALES

see **Bacteria**

EUKARYOTES

These organisms include all nucleated protozoa, most fungi and all algae except for cyanophyta (prokaryotic blue-green algae).

EUPHOTIC (PHOTIC) ZONE

The shallow zone in water through which sunlight can penetrate and in which photosynthetic activity can be carried out.

EUROPEAN FEDERATION OF NATIONAL ENGINEERING ASSOCIATIONS (FEANI)

In order that engineers may practice their profession freely among countries, professional engineering registration bodies in 22 European countries have formed a central registry, which has its headquarters in Paris. Persons who have met the licensing standards of their own country in terms of education and experience are eligible to be registered with FEANI and may use the title of European Engineer (EUR ING).

EUTROPHICATION

Aging of lakes and other water bodies. Nutrient enrichment is the fundamental cause of more rapid evolution into bogs and eventual disappearance. Associated with this problem are aquatic weeds and dense algal growth.

EVAPORATION

Loss of material from a water or other liquid surface. This occurs when the vapor pressure of the surface is greater than the vapor pressure of the overlying atmosphere.

EVAPOTRANSPIRATION

Lumping together the two processes of evaporation and transpiration. In practice, it is difficult to differentiate between the two, particularly in vegetated areas. see **Evaporation**. see **Transpiration**

EXCHANGE RATE (US)

The OSHA requirement that the permissible exposure time for a 90 dBA noise level be halved for each rise of 5 dBA.

EXCRETION

The three major routes of elimination of a toxicant are urine, feces and exhalation. Minor routes are hair, nails, saliva, skin, milk and sweat.

EXTRACELLULAR ENZYMES

These are excreted by the microbial cell and effects are outside the cell. see **Green-Stumpf theory**

EXTRACTIVE PROCEDURE TOXICITY (US)

A test designed to protect water supplies. A suspect waste is leached, as specified in the test method, and the leachate is analyzed for certain metals, pesticides and herbicides. If the concentration of any of these contaminants exceeds the allowable, the waste is given an EPA hazardous waste number.

EXTREME VALUE STATISTICS

see **Non-parametric statistics**

F

FABRIC FILTRATION

Dust and submicron particle collection efficiencies greater than 99% can be attained by filtration of dirty gas through porous fabric filters. The filter medium, such as in fossil fuel combustion plants, is contained in a structure known as a baghouse. Baghouses are common in many large industrial installations such as lead smelters.

FACULTATIVE ORGANISMS

These organisms can take oxygen for metabolic processes from dissolved molecular oxygen or combined molecular oxygen from ions such as nitrate, nitrite, sulfate, phosphate or borate. Objectionable conditions and odors associated with the anaerobic state commence at about the point of exhaustion of dissolved nitrate.

FACULTATIVE PHOTOTROPHS

Organisms which are able to utilize organic compounds in the dark and photosynthetically fix carbon dioxide in light.

FAO

see **Food and Agricultural Organization**

FATTY ACID

A carboxilic acid which may be branched or unbranched. The general formula is $R-(CH_2)_n-COOH$, where R is a hydrocarbon group.

FBC
see **Fluidized bed combustion**

FEANI
see **European Federation of National Engineering Associations**

FEANI INDEX
A list of the accredited engineering degree courses in each of the 22 member countries of the Association, with details of the educational system, composition, duration of study and awards of each engineering and technical school.

FEDERAL COAL LEASING AMENDMENTS
see **General Mining Act**

FEDERAL DRIVING CYCLE (US)
A standard developed for testing of motor vehicles for emissions. A short cycle has been developed for convenience in some testing. Involved are accelerations through several speed ranges, followed by deceleration and idle.

FEDERAL FACILITIES AGREEMENT
see **Compliance agreements**

FEDERAL FACILITIES COMPLIANCE ACT (US)
This 1992 law requires that all federal facilities be in compliance with all applicable federal and state hazardous waste laws. It waives federal immunity under these laws and allows imposition of fines and penalties.

FEDERAL INSECTICIDE, FUNGICIDE AND RODENTICIDE ACT (FIFRA) (US)
All new pesticides must be registered with the EPA. Pesticides previously registered must be reviewed with regard to potential adverse effects on human health and the environment.

FEDERAL LAND POLICY AND MANAGEMENT ACT (US)
The Bureau of Land Management of the Department of the Interior administers national resource lands under provisions of this Act. A significant policy set forth in the Act is that public lands should be retained under federal ownership unless disposal is in the national interest.

FEDERATION EUROPEENNE D'ASSOCIATIONS NATIONALES D'INGENIEURS
see **European Federation of National Engineering Associations**

FENTON'S REACTION
Reaction of ferrous iron and hydrogen peroxide to produce free radicals (OH).

FFA
see **Compliance agreements**

FFCA
see **Federal Facilities Compliance Act**

FICK'S LAW (OF DIFFUSION)
The change in concentration of a substance per unit time across an area is proportional to the molecular diffusivity of the substance and the concentration gradient.

FIFRA
see **Federal Insecticide, Fungicide and Rodenticide Act**

FILARIASIS
Caused by a variety of filarial nematode worms which colonize the lymphatic system and cause elephantiasis, a massive swelling of affected parts. River blindness, so called because it is contracted near water, results from infection of the retina and optic nerve.

FILTER FLY
see **Psychoda fly**

FILTRATION
The process of removing particles from a moving fluid by passage through small openings. In most filtering operations of environmental significance the mechanism is less that of physical straining than of compaction and adsorption onto surfaces of the filter medium.

FINDING OF NO SIGNIFICANT IMPACT (FONSI) (US)
When a federal agency takes the position that a proposed project or action will not have a significant impact on the environment, it must

prepare and circulate a mini-EIA (Environmental Impact Assessment) and a finding of no significant impact (FONSI).

FIRST ORDER REACTION

The rate of reaction is proportional to the amount of material not yet reacted. Examples are radioactive decay, BOD exertion and bacterial kill. see **Biochemical oxygen demand**. see **Chick's Law**. see **Zeno's paradox**

FISCHER-TROPSCH PROCESS

A method for synthesis of aliphatic compounds. A mixture of hydrogen and carbon monoxide is reacted in the presence of an iron or cobalt catalyst. Carbon monoxide is produced by partial oxidation of coal and wood-based fuels. The mixture of carbon monoxide and hydrogen is called synthetic gas (syngas). The environmental significance is the potential for reducing the need for oil in producing fuels.

FISH AND WILDLIFE COORDINATION ACT (US)

Federal agencies proposing water resource development projects are required to consult with the Fish and Wildlife Service with regard to conserving wildlife.

FISH AND WILDLIFE SERVICE (FWS) (US)

This is under the Department of Interior. FWS deals with preservation of fish and wildlife resources.

FISSION PRODUCTS

Radioactive nuclides, including primary and decay products, resulting from nuclear fission.

FJORD

Formed by glaciers and having a U-shaped cross-section, fjords often have a shallow depth (sill) at the mouth due to terminal glacial deposits. Basins inside the sill are usually quite deep.

FLOCCULATION

Addition of treatment chemicals forms precipitates and allows neutralization of surface charge on colloidal particles (coagulation). The small particles thus formed settle slowly. Flocculation is the gentle agitation which allows small particles to agglomerate, forming larger and more easily settling particles.

FLOOD (RESERVOIR) ROUTING

A computational method applied to determine the downstream effects of releasing flood flows into a lake or reservoir. This method is also applied to detention (retention) basins. see **Detention basin**

FLOW EQUALIZATION

Wastewater flows exhibit considerable variation over a 24 hour period. It is desirable to smooth out these flows in order to relieve hydraulic and organic peak loads. This is usually accomplished by holding basins.

FLUE GAS DESULFURIZATION

Removal of sulfur compounds from gaseous effluents. Methods are classified as regenerable or nonregenerable, depending on the end product (H_2SO_4, S).

FLUID

A substance which deforms continuously under any shear, no matter how slight. This is the reason that water flows downhill.

FLUIDIZED BED

A bed of particles held in suspension by a moving fluid, it is one in which no particle physically supports any other particle. Quicksand is an example of incipient fluidization.

FLUIDIZED BED COMBUSTION

Reaction of suspended coal particles with a flowing gas. The process is used for control of emissions from a coal fuel source. Beds are broadly classified as circulating and bubbling.

FOAM SEPARATION

This process removes refractory organics and heavy metals occurring at low concentrations by rising bubbles. The resulting surface foam is removed by mechanical means.

FOG

A colloidal dispersion of a liquid in air. Smog is an artificial fog produced near the ground by photochemical reactions in polluted atmospheres. see **Photochemical oxidants**. see **Smog**

FONSI
see **Finding of no significant impact**

FOOD ADDITIVES
Substances which function as coloring materials, flavor enhancers, shelf life extenders and in protection of food nutritional value.

FOOD AND AGRICULTURAL ORGANIZATION (FAO)
A United Nations agency for coordination of food, agriculture, fisheries and forestry programs over the world.

FOOD CHAIN
The series of interactions that occur among organisms in efforts to obtain food and energy. Because of energy loss in each transfer, most food chains involve only four or five links from beginning transfer of energy from primary producers through a series of consumers to the ultimate large consumer.

FOOD, DRUG AND COSMETIC ACT (US)
All food additives, color additives, drugs and cosmetics must be approved by the Food and Drug Administration before being offered for sale. No approval can be issued if substance is found to cause cancer in humans or laboratory animals.

FOOD WEBS
Interconnected food chains which may consist of grazing food chains involving direct consumption or detritus food chains which involve decomposition of dead matter.

FOREST AND RANGELAND RENEWABLE RESOURCES PLANNING ACT (US)
This Act consolidated certain functions of the US Forest Service and other functions were clarified. The Act was amended by the National Forest Management Act.

FORMALDEHYDE (CH$_2$O)
A colorless gas which is a strong irritant to the eyes and nasal tissues due to high water solubility.

FORMALIN
An aqueous solution of formaldehyde, it is an occupational lung allergen.

FOSSIL FUEL

Deposits of organic material formed under great pressure (coal, oil and gas) which can be burned. These are finite and will be replaced eventually by renewable sources.

FREE AVAILABLE CHLORINE

Hypochlorite ion (OCl^-) and undissociated hypochlorous acid result from reaction of chlorine with water. Distribution of the two species is a function of pH. The undissociated form is a strong disinfectant.

FREE RADICALS

Reactive intermediates such as an atom or fragment of a molecule with an impaired electron. OH radicals can initiate reactions which are responsible for oxidation of nitrogen oxide to nitrogen dioxide in the atmosphere. Free radicals are finding increased application in treatment of wastewater and wastewater sludges.

FRESH WATER

Water containing almost no chloride (0.03% chloride or less). see **Halinity**. see **Saline water**

FREUNDLICH ISOTHERM

An empirical logarithmic relationship describing adsorption from solution. An example is removal of tastes and odors from water supplies. see **Langmuir isotherm**

FRICTION FACTOR

see **Darcy-Weisbach equation**

FT

see **Fischer-Tropsch process**

FUEL CELL

A device for converting the chemical energy of hydrogen to electricity. By-products are water and heat. Of environmental significance is zero emission of carbon dioxide, sulfur oxides and nitrogen oxides. Fuel cells are considered promising in reduction of greenhouse gases. So far, applications have been restricted to smaller installations.

FUGITIVE EMISSIONS
Piles of material (sand, coal, etc.) in the open which can be carried by the wind.

FUMIGATION
The mixed layer in the atmosphere extends above a pollution source and the elevated polluted layer is mixed with ground air. This condition can extend a long distance from the source and bring about direct exposure at ground layer to an emission plume.

FUNGI
Multicellular or unicellular microorganisms with branched or filamentous structures and which utilize organic materials for food. Strictly speaking, bacteria are fusion fungi. Mold, in common usage, is taken to mean fungi. Lacking photosynthetic ability, fungi utilize organic material as a source of food and energy.

FUSION POWER
There are great reserves of deuterium in seawater. Reaction of deuterium with tritium or deuterium with deuterium could produce massive amounts of energy. The deuterium-tritium reaction involves confinement as a plasma at high temperature in a high strength magnetic field. The reaction will generate neutrons necessary for production of more tritium. A blanket of lithium can be circulated for generation of steam in a conventional generating plant.

FUZZY LOGIC
Introduced in the 1960s, fuzzy logic deals with uncertain or imprecise situations. Variables in fuzzy logic are described by linguistic symbols (small, medium, large) and these symbols are represented by fuzzy sets. Each set is characterized by a membership function which varies from 0 to 1. Fuzzy logic admits infinite logic levels to solve problems with uncertainties and is finding wider application in management, economics, medicine and systems. It allows for inclusion of operator experience.

FWS
see **Fish and Wildlife Service**

G

GAIA CONCEPT
An hypothesis which views biotic elements as attempting to moderate the local environment and optimize the physical and chemical environment. This has been a controversial concept but there is some supporting evidence for the views expressed.

GAMMA RAYS
These rays travel with the speed of light and are similar to X-rays but have shorter wavelengths and greater penetrating power. Shielding requires use of dense materials, such as lead.

GARBAGE
Animal and vegetable waste resulting from preparation of food.

GAS ABSORBER
A gaseous pollutant control device in which a soluble pollutant compound is removed from a gas stream by absorption into a liquid solvent.

GAS CHROMATOGRAPHY
An instrumental method of analysis in which a liquid sample is vaporized so that individual components can be identified separately.

GAS FLARING
The practice of burning waste gas from refineries. This is a significant source of greenhouse gases.

GASOHOL

The name given to mixtures of unleaded gasoline and ethanol (grain alcohol). This motor fuel is clean burning but is more expensive than conventional fuel. Negative aspects of gasohol are cost and removal of food producing areas from food production.

GEGEN ION

see **Counter ion**

GELLATION

In-situ solidification of a substance. This process is applied to oil and oil products to avoid spillage into the water after damage to a cargo compartment.

GENE

Unit of hereditable information occupying a fixed position on a chromosome. Genes direct protein synthesis.

GENERAL CIRCULATION MODEL

see **Global circulation model**

"GENERALLY RECOGNIZED AS SAFE" (US)

Terminology applied to a group of food additives accepted by the Food and Drug Administration. To be on the list, a substance had to be on the list before 1958 and meet specifications for safety. An additive can be deleted in light of new evidence. Cyclamate sweeteners and saccharin were formerly listed but were removed on the basis of animal tests. Substances introduced after 1958 must be tested individually for inclusion.

GENERAL MINING ACT (US)

This 1872 Act deals with location and patenting of claims to hardrock minerals. Quite out of date in some respects, it has not been a help in avoiding or mitigating environmental errors. However, environmental protection is one factor in awarding leases under provisions of the Mineral Leasing Act, the Outer Continental Shelf Lands Act, the Multiple Mineral Development Act, the Geothermal Steam Act and the Federal Coal Leasing Amendments.

GENETIC ENGINEERING

Techniques for deliberate modification of DNA in existing organisms and generation of completely new organisms.

GEOGRAPHICAL INFORMATION SYSTEM (GIS)

Developed originally by geographers, this computerized technique for gathering data, data reduction and application to real world problems is now finding much wider use.

GEOTHERMAL ENERGY

Temperature increases about 2°F per 100 feet (3.7°C/100 m) in rock and at a much greater rate near natural steam sources. Thermal energy from deep underground sources can be recovered by tapping natural steam or hot water or by fracturing rock and flooding. The energy in the rock can then be used to operate a conventional steam electricity generating plant. It is estimated that geothermal energy will be only 1–2% of the total energy consumption of the world.

GEOTHERMAL STEAM ACT

see **General Mining Act**

GIS

see **Geographical Information System**

GLOBAL CIRCULATION MODEL

Computer simulation of large-scale circulation of the atmosphere. This method is important in air pollution studies.

GLOBAL DIMMING

Pollutant particles in the atmosphere serve as nuclei for cloud formation. These clouds reflect sunlight back into space. As a result, since there is greater cloud formation than formerly, less sunlight reaches the surface. Global dimming actually cools the earth. Global warming affects the earth temperature to a greater degree than the cooling due to global dimming. Both global warming and global dimming affect weather patterns and one of the most serious effects is the change in the tropical rain belt. see **Greenhouse effect**. see **Tropical rain belt**

GLOBAL POSITIONING SYSTEM (GPS)

A system of satellites in orbit arranged so that, at any one time, at least four are visible from all points on the earth's surface.

GLOBAL WARMING

see **Greenhouse effect**

GLOBAL WARMING POTENTIAL (GWP)

A ranking of the absorptive capacity of the principal greenhouse gases. The atmospheric warming effect of each is compared to carbon dioxide, which is taken as one. The GWP of methane is 11, nitrous oxide is 270, CFC-11 is 4000 and CFC-12 is 8500.

GPS

see **Global positioning system**

GRAB SAMPLES

Single batch samples taken over a short period of time.

GRAHAM'S LAW

Rates of diffusion of gases are inversely proportional to the square roots of their densities.

GRAM MOLECULAR WEIGHT

The molecular weight, expressed in grams, of a compound.

GRAY (Gy)

see **Rad**

GRAY WATER

All domestic wastewater except that coming from water closets and latrines.

GREENFIELDS

Rural areas in danger of being converted to industrial areas. Conversion of greenfields usually involves construction of infrastructure while brownfields, in many cases, already have utility connections as well as transportation facilities. see **Brownfields**

GREENHOUSE EFFECT

Atmospheric concentrations of water vapor, carbon dioxide and methane affect the radiative heat balance and surface temperature of the earth. These gases, along with NO_x and chlorofluorocarbons, absorb infrared radiation and reemit it in all directions. Some of this returns to the surface, raising its temperature. Increases in CO_2 levels will, it is thought, raise the atmospheric temperature. see **Greenhouse gases**

GREENHOUSE GASES

The six gases and gas families that are covered under the Kyoto Protocol are carbon dioxide (CO_2), methane (CH_4), nitrous oxide (N_2O), hydrofluorocarbons (HFCs), perfluorocarbons (PFCs) and sulfur hexafluoride (SF_6).

GREEN-STUMPF THEORY

Microbial cells excrete extracellular enzymes which break down food in the water surrounding the cells. Smaller molecules resulting from this breakdown are then able to diffuse through the cell walls. The food is then metabolized within the cells with the aid of intracellular enzymes. see **Intracellular enzymes**

GREY LIST (EC)

Substances on this List are less toxic and less persistent than those on the Red List. They can be tolerated at low concentrations, depending on use. On this List are heavy metals, metalloids and their compounds, biocides and derivatives not on the Black List, organic compounds causing tastes and odors, toxic or persistent organosilicon compounds, phosphorous and inorganic phosphorous compounds, non-persistent oils and petroleum hydrocarbons, cyanide, fluorides and substances such as nitrates which affect the oxygen balance. see **Black List**. see **Red List**

GRIT

(1) Solid material, primarily sand, carried in flowing wastewater. It is necessary to remove these particles prior to treatment in order to protect pumps and maintain digester capacity. (2) Grit is the term applied to airborne particles with diameters of 76 microns or greater.

GRIT CHAMBER

A device at the head of a wastewater treatment plant for collecting particles, primarily sand, with a specific gravity of about 2.65. It is necessary to maintain a fairly narrow flow velocity in order to keep organic material from settling out. see **Detritus**

GRIT-POT

The base of an air sampling cyclone where the larger particles are deposited.

GROUNDWATER

Water found in consolidated or unconsolidated sedimentary deposits, fissures and limestone caverns. In the saturation zone of a water bearing stratum all of the voids are filled completely with water. Groundwater can be withdrawn by pumping from wells or can flow in a spring when the piezometric surface intersects the earth's surface. The greatest part of the world's water supply comes from groundwater.

GROWTH

Increase from within by taking in new products (assimilation) and incorporation into the organism's internal structure. Replication is by replication of DNA, the genetic blueprint. In viruses this may be RNA. Energy must be released in a controlled and usable manner by breaking down ATP.

GROWTH CURVE

For a microbial culture, progression is (1) the lag phase, where the organisms are becoming acclimatized, (2) the exponential phase where reproduction is at a constant rate, (3) the stationary phase in which the population is in equilibrium with the environment (endogenous), and (4) death.

GUMBEL DISTRIBUTION

see **Non–parametric statistics**

GWP

see **Global warming potential**

H

HABITATS DIRECTIVE (EU)

In 1992 the European Union launched a plan, known as Natura 2000, to develop conservation guidelines and a network of protected areas. The Habitats Directive, the tool for establishing the network, focuses on ecological communities rather on individual species. Member states were to implement management and protection plans by 2004. Some southern European member countries have large habitat areas which are considered ready for extensive development and this may cause future problems.

HAc

see **Acetic Acid**

HAGEN-POISEUELLE LAW

An expression for calculation of pressure loss in laminar flow, this was developed independently by a physicist and a physician who was studying blood flow.

HALDANE EQUATION

A description of inhibitory substrate reactions during nitrification, anaerobic sludge digestion, phenolic wastewater treatment and competition between carbon monoxide and oxygen for absorption onto hemoglobin.

HALF LIFE

The time, t, required for a substance to decay to one half the initial value. $t = 0.693/k$, where k is the reaction rate constant. k has the dimension of t^{-1}. see **Kinetics**

HALINITY

The degree to which water contains chloride. Sea water has about 35 000 mg/L of NaCl but this is about 22 000 mg/L as chloride. see **Brackish water**. see **Fresh water**. see **Saline water**

HALOALKANES

see **Halogenated organic compounds**

HALOGENATED HYDROCARBONS

Carbon and hydrogen compounds containing one or more of the elements fluorine, chlorine, bromine or iodine.

HALOGENATED ORGANIC COMPOUNDS

Called haloalkanes, these are generally toxic to humans. Some examples are chloroethane (ethyl chloride, C_2H_5Cl), chloroform (trichloromethane, $CHCl_3$), carbon tetrachloride (CCl_4) and chlorofluorocarbons (CFCs).

HARDNESS

The requirement for significant amounts of soap to produce lather or foam. Hardness is expressed as $CaCO_3$ and water with a hardness of 150 mg/L is considered hard. Principal hardness causing ions are Ca^{+2}, Mg^{+2}, Mn^{+2} and Fe^{+2}.

HARDY CROSS METHOD

An iterative (controlled trial and error) method for determining fluid (liquid and gas) flow in a complex pipe network. Flows are assumed in each pipe and head losses balanced by successive corrections or energy losses assumed and flows balanced. This method is analogous to Kirchkoff's Laws in electricity.

HAZARDOUS AIR POLLUTANTS (HAP) (US)

About 190 chemicals which are regulated by the Clean Air Act Amendments of 1990. Standards for the affected industries are found in the annually revised CFR 40, part 63.

HAZARDOUS AND SOLID WASTE AMENDMENTS (HSWA) (US)

These deal with protection of groundwater through leachate collection (double liners) and monitoring of underground tanks. Criteria for solid waste are upgraded.

HAZARDOUS ORGANIC NESHAP (NATIONAL EMISSIONS STANDARDS FOR HAZARDOUS AIR POLLUTANTS) (US)

Issued under the 1990 Clean Air Act Amendments, this requires that industries reduce hazardous emissions from facilities that make or use synthetic organic chemicals. Leaks from such sources as storage vessels, wastewater treatment units and transfer racks are covered. Affected are plants which produce or use 10 tons per year of one hazardous air pollutant (HAP) or 25 tons per year of multiple HAPs. see **Hazardous air pollutants**

HAZARDOUS POLLUTANTS

Pollutants, even if not widespread, which contribute to higher mortality rates in humans. see **Criteria pollutants**

HAZARDOUS WASTE

A solid waste that may cause or contribute significantly to serious health problems or death or that poses a substantial threat to human health or the environment when managed improperly. Included are materials which are ignitable, corrosive, reactive, explosive or toxic. Solid wastes include liquids and gases in containers and sludges.

HAZEN-WILLIAMS FORMULA

An observational expression for calculating flow velocity in a full flowing (pressure) pipe in terms of energy gradient, pipe material, pipe geometry and length.

HEALTH AND SAFETY AT WORK ACT (UK)

This 1974 Act deals with occupational health in the UK and is intended to secure health and safety for persons in the workplace, control hazards resulting from work and control emissions from the workplace. It requires that employers, as far as is reasonably practicable, provide a safe workplace.

HEALTH AND SAFETY COMMISSION (HSC) (UK)

Formed after the Health and Safety at Work Act of 1974, this semi-autonomous organization represents employers, local authorities and workers. Reporting is directly to the Secretary of State of the Department of Employment.

HEALTH AND SAFETY EXECUTIVE (HSE) (UK)

The HSE is the executive arm of the Health and Safety Commission and has responsibility for the Factory Inspectorate and Medical Advisory Service.

HEAT PUMP

A method of heating buildings and residences in which direct combustion of fossil fuel is not involved. Heat is absorbed from the surroundings or from waste liquid or gas streams. Refrigerant is evaporated by the external heat and vapor is heated. Heat from the vapor is next rejected to the space to be heated. There are four basic components of the system: (1) an evaporator where the liquid refrigerant at low pressure evaporates, removing latent heat from the environment; (2) a compressor for compressing low pressure gas; (3) a condenser in which high pressure vapor condenses, with attendant release of latent heat; and (4) an expansion valve for regulating the flow of liquid refrigerant from the high pressure side to the low pressure side.

HEAVY METALS

Included in this group are cadmium, chromium, copper, iron, lead, manganese, nickel and zinc. All except iron, manganese and zinc have adverse effects on plants and animals in moderate to elevated concentrations. Soluble iron and manganese in water cause staining of plumbing fittings when oxidized.

HELMINTHS

Parasitic worms which cause disease in humans and animals. Included are roundworms (nematodes), tapeworms (cestodes) and flatworms or flukes (trematodes). Among the diseases caused are trichinosis and schistosomasis.

HENRY'S LAW

The amount of gas that will dissolve in a given volume of liquid is directly proportional to the pressure exerted by the gas in contact with the liquid.

HEPATITIS

This disease is spread by drinking virus-contaminated water and eating shellfish which have accumulated viruses in the shell liquor. The virus passes unchanged through the stomach and lodges in the small intestine and moves to the liver, causing inflammation. Outbreaks are not usually fatal.

HEPTACHLOR

see **Organochlorine insecticides**

HERTZ

1 hertz (Hz) = 1 cycle/second.

HETEROCYCLIC COMPOUNDS

These have a ring structure with at least one ring carbon replaced. The most common contain oxygen or nitrogen. About half of all natural compounds are heterocyclic and can be aliphatic or aromatic.

HETEROTROPHIC

Other nourishing. An organism that uses other organisms, living or dead, as sources of carbon. In this component of an ecosystem, utilization, rearrangement and decomposition of complex materials are dominant.

HETEROTROPHS

These organisms need organic material as food and energy sources. Aerobes are heterotrophs which use dissolved oxygen. Anaerobes are those which oxidize organic matter in the absence of dissolved oxygen and facultative organisms are those which can carry on metabolism either with dissolved oxygen or can metabolize food using combined oxygen from other sources.

HEYWOOD DIAMETER

see **Equivalent diameter**

H-HORIZON

The uppermost part of the A-horizon in a soil.

HLB

see **Hydrophyllic-lipophyllic balance**

HLW
see **Radioactive waste**

HOLDING TANK
A closed container for holding sewage aboard watercraft until discharge to approved shore installations. Flush water is drawn into the system from overboard. These devices occupy a significant volume aboard a boat, can generate unpleasant odors and require support facilities ashore for emptying and cleaning. see **Incinerating toilets**. see **Maceration-disinfection devices**. see **Recirculating toilets**

HON
see **Hazardous Organic Neshap**

HSC
see **Health and Safety Commission**

HSWA
see **Hazardous and Solid Waste Amendments**

HUMAN AIRBORNE INFECTIONS
see **Airborne infections**

HUMUS
Decayed organic matter found in soils in the A-horizon.

HUMUS TANK
see **Secondary settling**

HUMAN WASTE (DEVELOPING COUNTRIES)
Only about one third of the populations of Developing Countries (Nations) are connected to sewer systems. Community sanitation in those areas without sewers must: (1) not be accessible to flies or animals; (2) cause odors or unsightly conditions; (3) be simple and inexpensive; (4) not contaminate surface water or groundwater; and (5) not cause soil contamination. The most common facilities are the bored hole latrine, box and can privy, chemical toilet, methane farming toilet, pit or vault privy, septic privy and soil (composting) toilet. The box and can privy is most commonly used in cities without sewerage systems. A scavenger service is necessary.

HYDRAULIC JUMP

A standing wave formed when a flow goes from a high velocity to a much lower velocity. This is a turbulent situation and can damage an unprotected stream bottom. The hydraulic jump is the incompressible flow analogy to Mach 1 in compressible flow.

HYDRAULIC LOADING RATE

The ratio of the mass of substrate applied per unit time to the reactor volume.

HYDRAULIC RADIUS

The ratio of the area through which flow passes to the wetted perimeter, where frictional losses occur. see **Darcy–Weisbach equation**. see **Hazen–Williams formula**. see **Manning equation**

HYDRAULIC RETENTION TIME

The ratio between reactor volume and volumetric flow rate.

HYDROCARBONS

Compounds containing carbon and hydrogen. In the presence of oxides of nitrogen and sunlight, hydrocarbons react to form photochemical oxidants.

HYDROFLUOROCARBONS

These have replaced CFCs in foam propellants and other former uses of CFCs. HFCs are much less damaging to the ozone layer than were CFCs.

HYDROGASIFICATION

Reaction of hydrogen with carbon bearing materials to produce methane-rich combustible gases. This means of biogas production is not suitable for small installations due to need for gas cleaning and skilled operation. see **Biogas**

HYDROGEN SULFIDE

A colorless gas with a rotten egg odor. At low concentrations H_2S has little effect on human health. It is quite toxic at high concentrations. Major sources include combustion of coal, oil and natural gas, petroleum refining, coke production, sulfur recovery and the Kraft process. see **Crown corrosion**. see **Mercaptans**

HYDROGRAPH

A plot of time versus instantaneous discharge. The area under the curve, minus base flow, is the volume of rainfall appearing as runoff. see **Base flow**. see **Detention basin**

HYDROLOGIC CYCLE

The path followed by water in Nature. Atmospheric water is deposited as rain, snow, sleet, hail and dew. Some of the water falling on the surface will run off directly to streams, rivers, lakes and oceans. The remainder will percolate into the ground. A portion of this will go into the water table and be stored. Some will be taken up by vegetation and eventually transpired back into the atmosphere. Some of the groundwater storage will be withdrawn for use and discharged to receiving waters. Waters on the surface of the ground and on the surfaces of water bodies will evaporate, thus completing the cycle.

HYDROLOGY

The science dealing with all phases of the transport of water among the atmosphere, the land surface and subsurfaces and the oceans.

HYDROPHILLIC-LIPOPHYLLIC BALANCE

A classification of surfactants for control of spilled oil, this is the ratio of the water compatible portion of the surfactant to the oil compatible portion.

HYDROPOWER

Production of electricity by use of flowing water to power generators. Generating plants can be placed in a river or water can be stored. Some countries, such as Switzerland, have used up almost all of the available hydrogenation capacity while others, such as the US, have developed only about 20%. There are environmental considerations, such as destruction or serious upset of the river life.

HYDROSPHERE

That portion of the earth's surface covered by water.

HYDROXY ACIDS

Some organic acids have OH groups attached to the molecule at points other than the carboxyl group. Thus, these compounds can act as acids or bases.

HYPOCHLOROUS ACID

Chlorine gas reacts almost instantaneously with water to form hypo-chlorous acid, HOCl. Undissociated hypochlorous acid is a much more effective disinfecting agent than is the hypochlorite ion, OCl^-. see **Disinfection**. see **Free available chlorine**

HYPOLIMNION

The stagnation zone in the deeper waters of a lake or reservoir.

I

ICD
see **International List of Diseases, Injuries and Causes of Death**

ICRP
see **International Commission on Radiological Protection**

IDA
see **International Bank for Reconstruction and Development**

ILW
see **Radioactive waste**

IMHOFF TANK
Also called an Emscher tank, this is a self-contained wastewater treatment unit for small communities and isolated installations not served by municipal systems. Operation is a combination of sedimentation and anaerobic digestion. Wastewater flows into the upper of the two chambers and solids settle into the lower section, where digestion takes place.

IMO
see **International Maritime Organization**

IMR
see **Infant mortality rate**

INCINERATING TOILETS

These are designed to destroy human waste, leaving only an ash residue. Energy sources are electricity, fuel oil and LPG. While not very efficient in terms of fuel usage, these units are practical for use in railroad cars and, to a lesser extent, watercraft. see **Holding tank**. see **Maceration-disinfection device**. see **Recirculating toilets**

INCINERATION

A controlled combustion process for conversion of solids, liquids and gaseous materials to gases and inert residue. The process requires careful design and continued operational monitoring.

INDEX CASE

The first case of an infectious disease.

INDIVIDUAL LIFETIME RISK

The unit risk value multiplied by the average exposure concentration over a lifetime of 70 years. see **Unit risk value**

INDIVIDUAL RISK

The product of the unit risk value and the ambient concentration of the pollutant under consideration. see **Unit risk value**

INDUSTRIAL ECOLOGY

This is an emerging field that deals with sustainability and focuses on industrial processes and related issues.

INDUSTRIAL HYGIENE

see **Occupational health**

INDUSTRIAL PRETREATMENT STANDARDS (US)

Some industrial wastes can upset or destroy biological wastewater treatment processes and some can pass unchanged through such systems. Under the National Pollutant Discharge Elimination System, industrial pretreatment standards are promulgated for such wastes before discharge to sewers leading to publicly owned treatment works.

INDUSTRIAL, SCIENTIFIC, MEDICAL FREQUENCIES (ISM) (US)

Microwave frequencies assigned by the Federal Communications Commission for industrial, scientific and medical applications.

INDUSTRIAL WASTE

These are strictly of industrial origin and must be differentiated from municipal waste. Some industrial wastes closely resemble sewage while others may contain toxic or non-biodegradable substances. Both BOD and COD are important in assessing the strength of waste. Newer plants have provision for waste treatment in the early design stage.

INDUSTRIAL WASTE SURVEY

These are conducted in order to develop information on wastes from an industry, sewer lines, waste routing and material balance.

INFANT MORTALITY RATE

A measure of the number of children born alive but not reaching the first birthday.

INFILTRATION

(1) Rain falling on a pervious surface and which percolates into the soil, satisfying soil moisture deficits. (2) Leakage of groundwater into loose joints of sewer pipes.

INFRASOUND

A sound wave with a frequency below the audible range.

INGESTION

Intake of food and energy.

INHERENT BIODEGRADABILITY

Potential for biodegradation. A negative test for biodegradation means that the substance can be assumed to be persistent in the environment. see **Winogradsky column**

INNOCENT LANDOWNER DEFENSE (US)

A section in the Superfund Amendments and Reauthorization Act which provides against Superfund cleanup liability for purchasers, lenders and financers of contaminated property who did not contribute to the contamination and who had no reason to believe that the property was contaminated when acquired.

INSOLATION

Solar radiation per unit area.

INSPECTORATE OF POLLUTION (HMIP) (UK)

This organization is responsible for monitoring air and water pollution and workplace cleanliness and safety. It has assumed the functions of the former Alkali and Clean Air Inspectorate.

INTEGRATED POLLUTION CONTROL (IPC) (UK)

A part of the Environmental Protection Act which treats all discharges to the environment as interrelated. see **Environmental Protection Act**

INTEGRATED RISK INFORMATION SYSTEM (IRIS) (US)

An electronic data base maintained by the US EPA that contains information on human health effects arising from exposure to hazardous pollutants.

INTEGRATED SAMPLES

see **Composite samples**

INTEGRATED WASTE MANAGEMENT

This approach for dealing with solid waste management involves reliance on a hierarchy of options from the most desirable to the least desirable. These are: (1) source reduction – limiting the amount and/or toxicity of wastes; (2) recycling – reuse of materials; (3) incineration – controlled burning; and (4) sanitary landfill – land disposal.

INTERFLOW

Lateral movement of water when it cannot percolate through a soil or rock formation.

INTERMITTENT SAND FILTERS

Beds of granular material underlain by graded gravel and collecting tiles used for BOD and suspended solids removal prior to soil infiltration. These units treat wastewater from individual homes.

INTERNATIONAL BANK FOR RECONSTRUCTION AND DEVELOPMENT (IBRD)

Together with the International Development Association, this forms the World Bank. Established by the UN in 1945, the function of IBRD is to assist economic development of member countries. Membership is restricted to member countries of the International Monetary Fund.

INTERNATIONAL COMMISSION ON RADIOLOGICAL PROTECTION (ICRP)

This Commission has recommended maximum permissible doses for workers in the nuclear industry and for the general population. The recommendations have received worldwide acceptance as the fundamental basis for international regulation.

INTERNATIONAL DEVELOPMENT ASSOCIATION

see **International Bank for Reconstruction and Development**

INTERNATIONAL LIST OF DISEASES, INJURIES AND CAUSES OF DEATH

Established through International Statistical Congresses in the nineteenth century, classifications of causes of death are now revised every ten years.

INTERNATIONAL MARITIME ORGANIZATION (IMO)

This United Nations organization has developed international agreements governing overboard discharge of contaminated ballast water by seagoing vessels. Enforcement is by national regulatory agencies of member countries. The Marine Environment Protection Committee deals with environmental matters, including transport of atomic substances by ships.

INTERNATIONAL STANDARDS ORGANIZATION (ISO)

A worldwide federation, founded in 1946, to promote development of global standards for manufacturing, trade and communication. ISO standards are voluntary but widespread adoption gives these standards almost mandatory status. ISO 14000 is intended to assist companies in performance improvement and keep environmental issues from becoming trade barriers. ISO 14001 is a proposed worldwide standard for environmental management systems. It differs from the European Union's Eco-audit in that it does not require an environmental report.

INTERNATIONAL SYSTEM (OF UNITS) (SI)

Fundamental quantities of this system are length (meter), time (second), mass (kilogram), electric current (ampere) temperature (Kelvin), luminous intensity (candela) and amount of substance (mole).

INTERNATIONAL UNION FOR CONSERVATION OF NATURE AND NATURAL RESOURCES (IUCN)

The primary purpose of this independent international organization is promotion of conservation of wildlife habitats and natural resources worldwide. The IUCN publishes its Red Data Books, which contain information on threatened species of plants and animals. Membership includes more than 400 government agencies and it works closely with UN agencies.

INTERNATIONAL WATER SUPPLY AND SANITATION DECADE

The United Nations declared 1981–90 as the period in which, it was hoped, the goal of supplying potable water to all peoples of the world might be attained. It was an admirable effort but fell far short of the goal.

INTRACELLULAR ENZYMES

These enzymes function on the protoplasm of the cell and act within the cell. see **Green–Stumpf theory**

INVERSION

see **Atmospheric inversion**

IODINE 131

This radioactive nuclide behaves the same as non–radioactive iodine and is concentrated in the thyroid gland, where it can cause cancer.

ION EXCHANGE

Displacement of one ion by another. Cations and anions can be exchanged between a liquid and a solid exchange medium. Cation, or base exchange, can be used to soften a water. Cation and anion exchange are both used in preparation of boiler feed water and industrial process water. Ion exchange media must have a large surface area per unit volume, high exchange capacity, be physically durable and relatively inexpensive, be capable of regeneration by inexpensive chemicals and must not discolor the water (free of color throw). see **Zeolites**

IONIC STRENGTH

A measure of the interionic effect resulting from electrical attraction and repulsion among various ions in solution. Ionic strength reduces

the apparent concentration of ions, leading to the concept of activity.
see **Activity**

IONIZATION
When a substance dissolves in water, it may break up into positive and
negative ions. Strong acids and strong bases may be considered 100%
ionized. Weak acids and weak bases are poorly ionized and are treated
by the appropriate ionization constant.

IONIZING RADIATION REGULATIONS (UK)
A set of regulations, under the Health and Safety at Work Act, dealing
with sealed and unsealed ionizing radiation sources.

IONOSPHERE
The part of the upper atmosphere, beginning at a height of 70–80 km,
which is highly ionized by solar ultraviolet radiation.

ISM
see **Industrial, scientific, medical frequencies**

ISO
see **International Standards Organization**

ISOMERISM
More than one organic compound can exist with the same empirical
formula. Structural formulae will differ.

ISOTHERM
see **Freundlich isotherm**. see **Langmuir isotherm**

ISOTOPES
These are atoms of the same substance but with different weights
(masses). A given element will have the same number of electrons and
protons. Isotopes are differentiated by the number of neutrons.

IUCN
see **International Union for Conservation of Nature and Natural
Resources**

J

JAR TESTS
Based on an analysis of raw water, it is possible to calculate the amounts of treatment chemicals required. However, in practice it is found more desirable to set up a series of beakers containing the raw water and add treatment chemicals in the region of the theoretical doses. The first in which floc particles appear while being gently agitated is taken as the optimum.

JET DROP PHENOMENON
This occurs in washbasins, bathtubs, toilet bowls and urinals when aerosols containing microorganisms are produced. Water-cooled dental drills, some humidifiers and poorly maintained ventilating system filters can also be sources of infectious disease.

JOULE
The work done when a force of 1 newton acts through a distance of 1 meter.

JUVENILE WATER
Water which originates from deep within the earth. Hot and often highly mineralized, it is important in groundwater recharge and is a source of geothermal energy.

K

KETONES

These compounds are prepared by oxidation of secondary alcohols. Acetone is the simplest ketone.

KINETICS

The study of the rate (velocity) of reactions.

KJELDAHL NITROGEN

Organically bound trinegative nitrogen. Ammonia nitrogen and organic nitrogen are reported as total kjeldahl nitrogen. Sources of the nitrogen are proteins and other substances contained in waste. Nitrite and nitrate are not included in total kjeldahl nitrogen.

KOZENTY EQUATION

This expression describes flow in granular materials. It describes head loss in a water filter in terms of the filter material. It applies in the laminar flow region.

KRAFT PROCESS

This involves cooking wood chips in sodium sulfide and sodium hydroxide to dissolve lignin. The strong odor from this process is due primarily to a mixture of hydrogen sulfide and dimethyl disulfide.

KREBS CYCLE

see **Citric acid cycle**

KYOTO PROTOCOL

This is an agreement made under the UN Framework Convention on Climate Change (UNFCCC). It came into force in 2005. Signers are committed to reduce emission of carbon dioxide and five other greenhouse gases or engage in emissions trading if they maintain or increase emission of these gases. Annex 1 countries are those developed signing countries which have accepted greenhouse gases reduction. Non-Annex 1 countries are those which do not have reduction obligations. Annex 1 countries must reduce their emissions 5% below 1990 levels by 2008–2012 but may meet their targets by purchasing emission credits. Non-Annex 1 countries may receive Carbon Credits when reduction projects are initiated. These Carbon Credits may then be sold to Annex 1 buyers.

L

LAG PHASE
see **Biochemical oxygen demand**

LAKE
A body of water occupying a depression in the earth's surface and which is completely surrounded by land. Lakes are divided into four zones: (1) the littoral zone extends from the shore to the deepest part which has rooted plants; (2) the pelagic zone extends from the littoral zone to the lake center; (3) the benthic, or profundal, zone is that part of the lake where light does not penetrate and there is no photosynthetic activity; and (4) the marginal zone is the immediate shoreline.

LAMBERT'S LAW
Intensity of light incident on an absorbing surface is decreased exponentially with increasing length of the light path. see **Beer's Law**

LAMINAR FLOW
Flow at low Reynolds number (less than 2000), the velocity profile is parabolic. Viscosity effects dominate. Groundwater flow and blood flow are examples of laminar flow. see **Reynolds number**

LANGMUIR ISOTHERM
A relationship between absorption of a gas and the partial pressure of a gas. see **Freundlich isotherm**

LA NINA

Cooler than average sea surface temperatures in the central and eastern tropical Pacific Ocean. When this occurs it has a significant effect on global weather patterns. Cooler than normal subsurface waters of the tropical Pacific are carried easterly by trade winds and upwelling occurs off the coasts of Ecuador and Peru. Sea surface temperatures fall. Effects on fisheries off the South American coast are benign. see **El Nino**. see **Upwelling**

LAPSE RATE

see **Adiabatic lapse rate**

LASER (LIGHT AMPLIFICATION BY STIMULATED EMISSION OF RADIATION)

A device that uses radiating properties of atoms or molecules to generate light of one or more discrete wavelengths. All waves are polarized, aligned and matching in phase. Lasing action can occur from ultraviolet to infrared wavelengths. The eye and, to a lesser extent, the skin are most susceptible to damage by lasers. Chronic and long term effects of lasers have not been fully explored.

LATENT PERIOD

The time between exposure to a carcinogen and development of the cancer.

LAW OF THE SEA (UNCLOS)

A number of conventions dealing with jurisdiction over ocean waters and use of ocean resources. Supported by Developing Nations, agreements reached were opposed by countries with vested interests in sea mining.

LDC

see **London Dumping Convention**

LD$_{50}$

see **Mean lethal dose**

LEACHATE

Precipitation percolating through a landfill will contain decomposed waste and bacteria. It is necessary to treat this water in order to avoid contamination of groundwater.

LEACHING PIT

see **Cesspool**

LEAD

A soft, malleable, bluish-gray heavy metal. Water and air are the principal means of lead into the body. Lead is no longer used as an anti-knock additive in motor fuel in the US but is used in some other countries. When water is soft, lead is taken into potable water, primarily from lead pipes and lead fittings (plumbosolvent). Between pH 7.0 and 9.5 there is a film formation of insoluble lead compounds which protect the lead surface. Young children can ingest lead from window sills and other painted surfaces. This is a particularly prevalent condition in poorer areas. Regulations now require removal of lead paint before a dwelling can be sold.

LEAK DETECTION AND REPAIR (LDAR)

A program implemented for compliance with regulations covering targeted chemicals in fugitive emissions. see **Fugitive emissions**

LEAST SQUARES

A procedure for fitting a linear equation to an array of data by minimizing the sum of the squares of the deviations of individual points from the line of best fit. The derived expression is the regression equation.

LEGIONNAIRES' DISEASE

A form of pneumonia first identified as a serious problem in 1976. The causative organism is *Legionella pnemophilia*. It grows slowly in an enriched, moist medium within a narrow pH range and is most active at around 104°F (40°C). The organism can survive for a year in tap water. The name of the malady is due to an outbreak at an American Legion convention in Philadelphia but it was determined that there were similar outbreaks previously. It is suspected that contaminated water in central air conditioning units or cooling towers can serve to disseminate *L. pneumophilia* in droplets into the surrounding atmosphere. The organism has also been associated with a syndrome called Pontiac fever, an influenza-like illness.

LEPTOSPIROSIS (WEIL'S DISEASE)

This disease is contracted by contact with urine of infected animals. Infection can affect the liver, kidneys and central nervous system and there is a typical jaundiced appearance. Prevention is by avoidance of water bodies likely to be polluted by animal waste.

LIFE TABLES
A means of showing mortality, by age. The beginning population is 1000 at age 0. Diminution by years is given until there are no survivors.

LIMNOLOGY
The study of freshwater lakes and streams.

LIPID
A class of molecules which are mostly straight chain. The chain length is between 12 and 24 carbon atoms. Biological membranes contain roughly 50% lipids and 50% proteins.

LIPID SOLUBLE
Fat soluble.

LIQUID BULK TANK VESSEL
The type of vessel which transports liquid cargos. It is the single greatest category of seaborne commerce and is the single largest potential source of oil pollution in terms of normal operation and casualties.

LIQUEFIED NATURAL GAS
Natural gas has a large volume per unit weight at ambient temperatures and must be carried in specially constructed tankers at −260°F (−162°C) in an independent tank or membrane liner.

LIQUEFIED PETROLEUM GAS
Butane, propane and ethylene are cooled to a low temperature (−46°C) and converted from the gaseous state to liquid. These are carried in specially constructed tank ships. There is great potential for explosions or spills. These gases can be carried at ambient temperature under high pressure in containers which are designed according to the ASME Code for Unfired Vessels.

LITTER
Street refuse.

LITTORAL ZONE
The sea bottom between high and low tides or the shore zone of a lake.

LLW
see **Radioactive waste**

LNG
see **Liquified natural gas**

LOCALLY UNACCEPTABLE LAND USE (LULU)

The term has been coined to describe local opposition to such new installations as sewage treatment plants, transfer stations, compost facilities, incinerators and recycling facilities.

LOG-GROWTH PHASE

In growth of microorganisms, where there is excess food, the rate of metabolism and growth will be controlled by the ability of the microorganisms to process the substrate (food). Here, microorganisms are removing organic material at their maximum rate. This phase is of great interest in waste treatment. Food (waste) concentration must be high and it is difficult to produce a stable effluent.

LONDON DUMPING CONVENTION (LDC)

An international convention dealing with sea disposal of wastes. In 1983 the LDC passed a resolution imposing an international moratorium on sea-dumping of solid radioactive wastes. While not binding on signatories, there was no further dumping of low level and intermediate level wastes. In 1993, a binding resolution against sea-dumping of radioactive waste was passed. There will be a scientific evaluation of the ban about the year 2019.

LONG RANGE TRANS-BOUNDARY AIR POLLUTION (LRTAP)

Use of information about inversions, topography, meteorology and climatology to develop a mathematical model for prediction of acid deposition in the model area. Dimensions are of the order of 1000 km or more.

LONG-RANGE TRANSPORT MODEL (LRT)
see **Long-range trans-boundary air pollution**

LPG
see **Liquefied petroleum gas**

LRT
see **Long-range transport model**

LRTAP

see **Long range trans-boundary air pollution**

LULU

see **Locally unacceptable land use**

LYSIS

When microbial cells die, cell contents are released to the surrounding water, adding to the extracellular matrix.

M

MAB

see **Man and the Biosphere Program**

MACERATION-DISINFECTION DEVICE

Overboard discharge of untreated human waste from watercraft is no longer permitted. One means of dealing with this problem is to grind (macerate) the waste into small particles and mix with a disinfectant, usually hypochlorite, for a short period before overboard discharge.

MACERATOR-CHLORINATOR

see **Maceration–disinfection device**

MACROMOLECULES

see **Polymers**

MAGMATIC WATER

see **Juvenile water**

MAGNETOHYRODYNAMICS

Hot gases, seeded with potassium or cesium to make them conductive, are expanded at high velocity in a magnetic field. Direct current produced in the moving conductive gas flows to electrodes located in the gas channels. This method of energy production has a thermal efficiency of 50–60% when operated in the range of 2200°C. There are many practical problems associated with construction of a full scale

MHD plant, the foremost of which is material suitable for extended operation at the high temperatures of the process.

MAJOR SOURCES (US)
Those whose emissions quantities are large enough to have a dominant role in the air pollution status of an area. Depending on the air quality of a region, emissions as low as 10 tons per year could constitute a major source.

MALARIA
This is caused by the pathogenic protozoan *Plasmodium*. The vector is the female anopheles mosquito. Four species of *Plasmodium* infect humans and infections can range from mild, but chronic, to fatal. Prevention is by eradication of the breeding of the vector in stagnant water. Malaria is thought to be responsible, directly or indirectly, for more deaths than any other disease.

MALATHION
see **Organophosphorous**

MALTHUSIAN
The nineteenth century thesis of Thomas Malthus was that population increases exponentially while food production increases arithmetically. Food production has not followed his model but there is controversy over the limit to growth concepts.

MAN AND THE BIOSPHERE PROGRAM (MAB)
A UNESCO project to promote scientific rather than subjective grounds for long-term conservation of critical ecosystems.

MANGROVE
A formation of trees and shrubs along tropic and subtropical coasts. Distribution is largely limited to the tidal zone and usually to narrow strips of shallow coastal waters, deltas, estuaries and lagoons.

MANNING EQUATION
An expression for calculation of flow velocity in open (non-pressure) channel in terms of the energy gradient, channel geometry and empirical roughness factor n. n is a function of the channel material.

MARINE ENVIRONMENT PROTECTION COMMITTEE
see **International Maritime Organization**

MARINE PROTECTION, RESEARCH AND SANCTUARIES ACT (US)
This Act, commonly called the Ocean Dumping Act, prohibits all transportation of material from the US for the purpose of ocean dumping except as authorized by permit. Sewage and oil are covered under the Clean Water Act.

MARINE SANITATION DEVICES
see **Holding tank**. see **Incinerating toilets**. see **Maceration-disinfection**. see **Recirculating toilets**

MARPOL CONVENTION FOR THE PREVENTION OF POLLUTION FROM SHIPS (IMO)
The results of the international accord include constraints on maximum cargo tank size and the requirement of segregated ballast tanks on all new tanker construction larger than 20 000 deadweight tons (18 144 metric tonnes) capacity.

MARTIN DIAMETER
The length of a chord dividing a particle into two equal areas. This is significant in aerosol characterization.

MA$_7$CD$_{10}$
Critical stream flow, defined as the minimum average 7 consecutive day flows with a recurrence interval of ten years. This is a basis for evaluating dilution for waste discharge assimilation.

MATC
see **Maximum acceptable toxicant concentration**

MATERIAL SAFETY DATA SHEETS (US)
Facilities that, under the Federal Occupational Safety and Health Act, are required to submit data annually on amounts of hazardous chemicals present on the premises to local and state authorities must supply "material safety data sheets" which describe the substances on hand. In addition, these facilities are obligated to submit "emergency and hazardous chemical inventory forms."

MAXIMUM ACCEPTABLE TOXICANT CONCENTRATION (MATC)

The highest continuous concentration of a toxicant which does not significantly decrease laboratory fish survival.

MAXIMUM PERMISSIBLE DOSE (MPD)

The amount of radiation to which workers may be exposed in one year. This dosage is under periodic review and can change in light of new knowledge.

MEAN CELL RETENTION TIME

see **Sludge age**

MEAN HOLDING TIME

see **Hydraulic retention time**

MEAN LETHAL DOSE

The dose required to kill one half of a target population.

MERCAPTANS

Also known as thioalcohols, these are aliphatic compounds containing sulfur. The structure is similar to alcohols but oxygen is replaced by sulfur. These compounds have offensive odors. The Kraft, or sulfate, process for pulping wood produces mercaptans. Mercaptans have been found to be toxic to fish.

MERCURY

A silvery white metal, liquid at ambient temperatures. The chief mineral source is cinnabar (HgS). Mercury may be absorbed through the skin, respiratory tract and digestive tract and is mostly excreted through the urine. The metal is a severe protoplasmic poison which causes both acute reactions and chronic illness. Metallic mercury discharged into Lake St Clair was found to be taken up after metabolism to fish and, ultimately, to humans. Methylmercury was found to cause Minamata disease in Japan. Mercury concentrations are usually expressed in micrograms per liter. see **Minamata disease**

MERV

see **Minimum efficiency reporting value**

MESOLIMNION
In a lake or reservoir, the transition zone between the upper (epilimnion) zone and the stagnant bottom (hypolimnion) zone.

MESOTHELIOMA
A diffuse malignancy of the lining of the chest cavity or the abdominal lining as a result of continued exposure to asbestos. The time between exposure and onset of the disease can vary from 10 to 35 years.

MESOTROPHIC
Medium productivity in an aquatic system.

METABOLIC REACTIONS
In wastewater treatment these are oxidation, synthesis and endogenous respiration.

METALLOIDS
A class of metals which are neither metals or non-metals. Included are boron, silicon, germanium, arsenic, antimony, tellurium and polonium.

METEOROLOGY
The study of weather and climate. Synoptic meteorology is concerned with weather forecasting.

METHANE
The simplest of the homologous series C_nH_{2n+2}. This gas (fire damp) is easily burned and is a major constituent of wastewater sludge digester gas.

METHANE FARMING TOILETS
see **Human waste**

METHANE HYDRATES
Massive amounts of methane are held in solution in ice caps. There is fear that further melting of this ice will release this greenhouse gas.

METHEMOGLOBINEMIA
A condition in which infants, usually less than two months old, fed on formulae prepared with water having high nitrate concentration, became

cyanotic. Nitrate is reduced in the intestine to nitrite which combines with hemoglobin, rendering it incapable of carrying oxygen. The normal oxygen carrier is oxyhemoglobin (HBO_2).

METHYLATION
Microorganisms can act on heavy metals and metalloids to produce alkyl compounds which are toxic and bioaccumulative.

METHYLENE BLUE
This dye remains blue when there is dissolved oxygen present in water. When all DO is exhausted, the blue color disappears. Decolorization of methylene blue is a routine test in operation of a wastewater treatment plant. Methylene blue was used as a marker in HEPA filter testing until it was replaced by the sodium flame test. Air filters were quoted as having an MB (methylene blue) efficiency.

METHYLMERCURY
Microorganisms can transform metallic mercury, phenylmercury and alkylmercury into methylmercury. This form is assimilated and accumulated by marine life up the food chain from plankton to large fish, birds and mammals. Methylation continues as long as there is a mercury source and organisms to process it.

METHYL TERT-BUTYL ETHER (MTBE)
The most commonly used fuel oxygenate for reducing carbon monoxide and unburned hydrocarbons in automobile exhaust. This additive raises the octane level of the fuel, increasing combustion temperature and engine efficiency. Other related oxygenates are ethyl tert-butyl ether (ETBE), tert-amyl methyl ether (TAME) and tert-amyl ethyl ether (TAEE).

M/F RATIO
The ratio of microorganisms to food. This is important in operation of a biological wastewater treatment process. A low M/F ratio indicates that microorganisms are in the log-growth phase while a large M/F ratio indicates endogenous respiration.

MHD
see **Magnetohydrodynamics**

MICHELIS-MENTON REACTION

A complex kinetics expression describing the rate of biological waste treatment, it was first used to describe enzyme reactions. Bacterial decomposition involves a series of enzyme catalyzed steps and the Michaelis-Menton reaction can be applied to bacterial growth and waste decomposition kinetics. see **Monod equation**

MICROBIAL INFALLIBILITY PRINCIPLE

Whatever Man or Nature can make, microorganisms can degrade.

MICROBIAL INSECTICIDES

Derived from microbial sources, these are highly specific and do not leave persisting residues, as is the case of chlorinated pesticides.

MICROBIOLOGY

The scientific study of microorganisms, which includes algae, bacteria, molds, protozoa, yeasts, viruses and rickettsiae.

MICROSCREEN

see **Microstrainer**

MICROSOMAL ENZYMES

Originating in the liver, these are involved in metabolism of drugs and compounds foreign to the organism. Lipid-soluble materials are converted by oxidation and reduction processes to less lipid-soluble substances which are more easily excreted by the kidneys.

MICROSTRAINER

Also called microscreens, this device is used to remove suspended solids from secondary effluents. Filter fabric on the periphery has openings in the range of 25 microns. Water enters the open end of the filter drum and flows out through the rotating screen. Backwash requirements are of the order of 2%.

MICROWAVE RADIATION

Wavelengths vary from about 1 nm to 10 m. The region between 10 MHz and infrared is the radiofrequency (RF) region. Microwave applications have become common in everyday life. Biological effects are thought to be primarily on the eye and skin. Performance standards have been promulgated for microwave ovens and specify a level of 1 nW/cm^2 at any point 5 cm or more from external oven surfaces. Cardiac pacemakers may have function compromised by microwave radiation. There is continued controversy over effects of radar waves on humans.

MINAMATA DISEASE

The name of this malady is taken from the Japanese town in which it occurred. Inorganic mercury was discharged in the effluent of a local industrial plant. The mercury was converted by marine organisms to lipid-soluble methylmercury, which was taken up in the food chain to the fish which formed the staple local diet. In a four year period in the 1950s 43 deaths and about 70 serious illnesses were acknowledged by the authorities. Some unofficial estimates put the death total as high as 800. In 1989 two former company executives were given prison terms for the 1950s pollution.

MINERAL LEASING ACT

see **General Mining Act**

MINIMUM EFFICIENCY REPORTING VALUE (MERV)

This is a system based on a standard for rating filters for removal of particulate matter from the air. The standard was published by the American Society of Heating, Refrigerating and Air Conditioning Engineers (ASHRAE). Values of 17–20 are associated with high-efficiency particulate air (HEPA) filters.

MINOR SOURCES

Air pollution sources which cannot be classified easily. They may be stationary or mobile. Examples are combustion for heating of residences and commercial buildings, service industries, animal and food processing and handling of paint and other surface coverings.

MIXED LAYER

In air pollution, the distance from ground level to the bottom of the inversion layer.

MIXED LIQUOR SUSPENDED SOLIDS (MLSS)

The concentration of biomass in an activated sludge aeration tank. see **Activated sludge**

MIXED LIQUOR VOLATILE SUSPENDED SOLIDS

This is a closer estimate of the viable organisms in activated sludge than mixed liquor suspended solids due to the presence of inorganic solids.

MIXED MEDIA FILTER

Sand, when used alone in water filtration, is graded hydraulically during backwashing. As a result, the smallest particles are at the top of the filter and most of the floc is removed in the upper layer and clogging occurs rapidly. To overcome this problem, a filter composed of coarse coal (Anthrafilt), a layer of sand and a layer of garnet is constructed. Coal is lighter than sand and sand is less dense than garnet. Pore size is greater toward the top in this arrangement. see **Rapid sand filter**

MIXED VALENCE

The same chemical element is found in two oxidation states in the same molecule.

MLSS

see **Mixed liquor suspended solids**

MLVSS

see **Activated sludge**. see **Mixed liquor volatile suspended solids**

MODELING

Mathematical description and representation of a phenomenon or process.

MOGDEN FORMULA (UK)

This expression allows calculation of the charges to the producer of an industrial pollutant discharged to a municipal treatment facility. see **Polluter Pays Principle**

MOLE

The actual concentration of the substance present, expressed in g/L, divided by the molecular weight of the compound. Concentration is expressed in moles per liter (mol/L).

MOLECULAR SIEVE

see **Zeolites**

MONOD EQUATION

This expression relates specific growth rate of microorganisms and substrate (food) concentration.

MONOSACCHARIDE

Naturally occurring monomers (sugars) are tetrose, pentose and hexose. The general formula is $C_nH_{2n}O_n$ but, due to isomerism, there are many monosaccharides.

MONOSODIUM GLUTAMATE (MSG)

A widely used flavor enhancer. Excessive intake of MSG has been linked to Kwok's disease (Chinese restaurant syndrome), a tightening of the facial muscles. Some countries have placed limits on MSG in foods.

MONTREAL PROTOCOL

As a result of this 2005 Conference on Climate Change, Kyoto Protocol signatories agreed to extend the agreement beyond the first commitment period ending in 2012. New emissions targets were not set but a large group of countries has agreed on non-binding talks on long-term measures.

MOODY DIAGRAM

A graphical representation of the Darcy-Weisbach friction factor f in terms of Reynolds number and relative roughness. It is used in iterative solutions of pipe flow problems.

MOST PROBABLE NUMBER (MPN)

This is a somewhat crude quantitative method for enumerating coliform concentrations. It is a statistical means of determining the most probable number of organisms growing in several tubes in a series of dilutions. The usual method is inoculation of tubes of lactose broth and 24 hour incubation at $35°C$. Gas production is considered presumptive evidence of coliforms but false readings can be given by *A. aerogenes*. Standard Methods gives tables for evaluation of the test. The membrane filter test is faster and more accurate.

MOVING-BED FILTER

A continuous sand filter moving countercurrent to influent wastewater. Solids are retained on and within the filter. Solids are scrubbed from the filter medium and discharged as waste sludge and the washed media fed back to the bed.

MPD
see **Maximum permissible dose**

MPN
see **Most probable number**

MSDS
see **Material safety data sheets**

MSG
see **Monosodium glutamate**

MSW
see **Solid waste**

MTBE
see **Methyl tert-butyl ether**. see **Reformulated gasoline**

MULTIMEDIA FILTRATION
see **Mixed media filter**

MULTIPLE MINERAL DEVELOPMENT ACT
see **General Mining Act**

MULTIPLE USE AND SUSTAINABLE YIELD ACT (US)
This Act directs that planning and management activities of the public domain are to be taken to protect environmental, ecological, scientific, scenic, historical, air and water resources.

MUNICIPAL SOLID WASTE
see **Solid waste**

MUNICIPAL WASTEWATER
see **Sewage**

MUSKINGUM METHOD
A channel routing method used in hydrology to determine river channel responses to flood flows.

MUTAGENICITY

Chemical transformation of a gene (point mutation) or rearrangement of a chromosome (chromosome aberration). Almost all mutations are harmful or neutral.

MYCOPLASMA

Prokaryotes which overlap viruses in size (100–300 nm). They do not have one typical shape but can appear coceoid, filamentous or highly branched.

N

NAAQS
see **National Ambient Air Quality Standards**

NADP/NTN
see **National Atmospheric Deposition Program/National Trends Network**

NARCOSIS
Central nervous system disorder.

NASOPHARYNX
The portion of the respiratory tract from the mouth to the epiglottis and larynx.

NATIONAL AMBIENT AIR QUALITY STANDARDS (NAAQS) (US)
Primary standards designed to protect the public health of the most susceptible groups in the population. Secondary standards have been set to protect the public welfare, including damage to plants and materials, and esthetic effects, such as reduction in visibility.

NATIONAL ATMOSPHERIC DEPOSITION PROGRAM/NATIONAL TRENDS NETWORK (NADP/NTN) (US)
A system of more than 200 air sampling sites for gathering precipitation chemistry data.

NATIONAL EMISSIONS STANDARDS FOR HAZARDOUS AIR POLLUTANTS (NESHAP) (US)

Sources which emit asbestos, beryllium, mercury, vinyl chloride, radio-nuclides, or any of 189 listed compounds.

NATIONAL ENVIRONMENTAL POLICY ACT (NEPA) (US)

"Environmental full disclosure law." Federal agencies which propose an action which may have a significant effect on the human environment are required to prepare and circulate an Environmental Impact Statement (EIS). The EIS must discuss the proposed action, reasonable alternatives and the commitment of resources. An adverse EIS is not, in itself, reason for rejection of a proposed action. Often an EIS becomes a massive justification for a decision already made and is not a reflection of a legitimate effort to avoid adverse environmental effects. Many other government entities at all levels now have requirements for environmental impact assessments of proposed projects. see **Environmental Impact Statement**

NATIONAL FOREST MANAGEMENT ACT

see **Federal Land Policy and Management Act. see Forest and Rangeland Renewable Resources Planning Act**

NATIONAL HISTORIC PRESERVATION ACT (US)

This legislation established the National Register of Historic Places. Federal agencies must consult with the Advisory Council on Historic Preservation whenever proposed federal projects might have adverse effects on historic or archaeological sites listed or eligible for listing on the National Register.

NATIONAL INSTITUTE FOR OCCUPATIONAL SAFETY AND HEALTH (NIOSH) (US)

This was established under provisions of the Occupational Safety and Health Act of 1970 and is responsible for formulating new or improved occupational safety and health standards for the workplace. It has a function similar to the health program under the 1969 Federal Coal Mine Safety and Health Act.

NATIONAL MONUMENTS

see **Antiquities Act**

NATIONAL OCEANOGRAPHIC AND ATMOSPHERIC ADMINISTRATION (NOAA) (US)

An Agency, under the Department of Commerce, comprised of the National Ocean Survey (formerly the Coast and Geodetic Survey), National Weather Service (formerly the Weather Bureau), National Marine Fisheries Service, Environmental Data Service, National Satellite Service, Environmental Research Laboratories, Office of Sea Grants and the Office of Coastal Management.

NATIONAL POLLUTANT DISCHARGE ELIMINATION SYSTEM (NPDES) (US)

Operators discharging pollutants and potential discharge operators must apply for an NPDES permit. The application must contain accurate discharge information. NPDES permits contain effluent limitations, compliance schedules and monitoring and reporting schedules.

NATIONAL PRIORITY LIST (US)

Under Superfund (CERCLA), a list of hazardous waste sites is prepared. Remedial feasibility studies for these sites are mandated.

NATIONAL RADIATION PROTECTION BOARD (NRPB) (UK)

Created under the Radiological Protection Act of 1970, the Board provides monitoring of ionizing radiation and advises on problems.

NATIONAL RIVERS AUTHORITY (NRA) (UK)

Established under the Water Act of 1989, this Authority deals with water quality, pollution monitoring, management of water sources, drainage and flood control.

NATIONAL WILDLIFE REFUGES SYSTEM ADMINISTRATION ACT (US)

This statute deals with National Park Service administration of National Wildlife Refuges. Hunting, fishing and other public recreational activities are permitted if they are compatible with the conservation purposes.

NATURA 2000

see **Habitats Directive**

NATURAL GAS

Natural gas resources are classified as nonassociated (existing alone), associated (existing in a gas cap underlain by oil) or dissolved in oil.

Ultimate recovery of natural gas in the world is estimated to be
8000–12 000 × 10^{12} ft³ (2.3–3.4 × 10^{14} m³) and 1200 × 10^{12} ft³
(3.4 × 10^{13} m³) in the US Natural gas delivered to the consumer is
primarily methane. Gas from the field contains hydrocarbons of the
series C_nH_{2n+2}, natural gas liquids, hydrogen sulfide and, at times,
helium.

NATURAL RESOURCE DAMAGES (US)

A provision of CERCLA (Superfund) allows designated state and
federal agencies to recover damages for injury, loss or destruction of
natural resources caused by release of hazardous substances.

NATURAL WASTEWATER TREATMENT SYSTEMS

These are divided into two major types: terrestrial and aquatic. The
terrestrial (land application) are slow-rate irrigation, rapid-infiltration-
percolation and overland flow. The aquatic are pond systems, anaerobic
ponds and facultative ponds. Both types depend on physical and chemical responses and unique biological components. see **Constructed
wetlands**

NECROSIS
Death.

NEMATODES
see **Helminths**

NEPA
see **National Environmental Policy Act**

NESCAUM
see **Northeast States for Coordinated Air Use Management**

NESHAP
see **National Emissions Standards for Hazardous Air Pollutants**

NESSLER TUBES

These are also called color comparison tubes and use is based on Beer's
Law. A series of tubes of equal length is set up so that slight differences
in color produced by analytical methods can be determined visually. It
is necessary to set up standard concentrations. These tubes now find

limited use and are most applicable to determination of substances at low concentration. see **Beer's Law**. see **Lambert's Law**

NEUTRON
An elementary particle with closely the same mass as a proton but which carries no charge.

NEW SOURCE PERFORMANCE STANDARDS (NSPS) (US)
Air pollutant sources for which national emissions standards have been established.

NEWTON
A force which will impart an acceleration of 1 m/s^2 to a mass of 1 kilogram.

NICKEL
A grayish-white metal. Major pollutant sources are nickel dust and vapors. Effects of nickel include carcinoma of the mucous membranes of the nose, associated air sinuses and lungs. Nickel affects DNA replication.

NIGHT SOIL
Human feces collected for use as fertilizer.

NIMBY
see **"Not in my backyard"**

NIOSH
see **National Institute for Occupational Health and Safety**

NITRATE (NO_3^-)
The highest oxidation state of nitrogen to be found in water and waste-water. Nitrite (NO_2^-) is oxidized by the *Nitrobacter* group to nitrate. The presence of a high nitrate concentration in wastewater effluent is considered an indication of a stable effluent. However, it is thought that nitrate can contribute to eutrophication. Nitrate in well water supplies has been shown to cause methemoglobinemia in infants. see **Methemoglobinemia**

NITRIC OXIDE (NO)
A colorless, odorless, tasteless and relatively non-toxic and non-irritating gas. Sources include anaerobic biological processes, photochemical

destruction of nitrogen compounds in the stratosphere and combustion processes.

NITRIFYING BACTERIA

These organisms convert ammonia to nitrite and nitrate. Nitroso-bacteria convert ammonia to nitrite and Nitro-bacteria convert nitrite to nitrate.

NITRILOTRIACETIC ACID (NTA)

A substance substituted for phosphates in detergents. Phosphates have been implicated in eutrophication. see **Synthetic detergents**

NITRITE (NO_2^-)

This ion appears in the intermediate stage of wastewater stabilization. Excess ammonia is oxidized by microorganisms to nitrite. Concentrations are usually less than 1 mg/L NO_2^- –N in effluents. In surface and groundwater supplies concentrations will be less than 0.1 mg/L NO_2^-–N.

NITROGEN

Nitrogen compounds can occur in the environment from natural and anthropogenic sources. The total pathway can be from urea, H_2NCONH_2, under aerobic and anaerobic conditions, to ammonia. Under aerobic conditions ammonia is converted by microorganisms to nitrite and nitrate. see **Ammonia**

NITROGEN DIOXIDE (NO_2)

A colored gas with a pungent irritating odor. It is a light yellowish orange at low concentrations and reddish brown at high concentrations. Relatively toxic, it is corrosive due to a rapid oxidation rate. Major NO to NO_2 processes are photochemical.

NITROGENOUS OXYGEN DEMAND

The later stages of biochemical oxygen demand exertion. The carbonaceous phase is essentially complete and now ammonia and nitrites are being oxidized to nitrates.

NITROGEN OXIDES (NO_x)

The term nitrogen oxides and the formula NO_x refer to mixtures of nitrogen oxide (NO) and nitrogen dioxide (NO_2), which are not distinguished as to source or effects.

NITROUS OXIDE (N₂O)

A colorless, slightly sweet non-toxic gas. Known as "laughing gas," it is an anesthetic.

NMOC

see **Non Methane Organic Carbon**

NOAA

see **National Oceanographic and Atmospheric Administration**

NOISE

Unwanted sound. A group of sound waves arriving simultaneously from a number of sources and having a wide range of frequencies.

NOISE CONTROL ACT (US)

This 1972 Act sets standards for industrial noise exposures and guidelines for hearing protection. OSHA, in 1981 and 1983, set specific criteria for exposure.

NONATTAINMENT AREAS (US)

Under the Clean Air Act, new sources of air pollution, emissions of hazardous air pollutants, emissions limits and transportation control plans are regulated in areas where federal standards are exceeded.

NON-IONIZING RADIATION

Electromagnetic radiation which does not cause ionization in biological systems and which have photon energies less than 10–12 eV. Sources are infrared, visible, ultraviolet, laser and microwave radiations. The eye is the organ most at risk. see **Electromagnetic spectrum**

NON METHANE ORGANIC CARBON (NMOC)

A parameter used to describe normal air emissions in an urban area. NMOC/NO$_x$ gives the normal emissions and is the starting point for evaluating control alternatives.

NON-MIX FLOW

see **Plug flow**

NON-PARAMETRIC STATISTICS

Non-normal statistics. Many natural phenomena do not follow a normal distribution. A widely used means of predicting infrequent events is the extreme value, or Gumbel, distribution. Infrequent occurrences and associated return periods often must be predicted on the basis of limited data and it is necessary to fit existing records to a distribution. see **Normal distribution**

NONPOINT SOURCE
A man-made or man-induced pollutant source that brings about alteration of the biological, chemical, physical or radiological integrity of the receiving water.

NON-RENEWABLE ENERGY
see **Fossil fuels**

NORDWIJK DECLARATION (ON CLIMATE CHANGE)
A Convention dealing with world climate change signed by 67 countries and 11 International Commissions and agencies. It deals with reduction of greenhouse gas production and increase of sinks of these gases. The 1989 Nordwijk Declaration followed the 1985 Vienna Convention on Protection of the Ozone Layer and the 1987 Montreal Protocol on Substances that Deplete the Ozone Layer.

NORMAL DISTRIBUTION
This bell shaped curve applies when the distribution can be assumed to result from a large number of independent variables. It is described fully by the mean and standard deviation. Probabilities of occurrence of events are proportional to areas under the normal curve. It may be convenient to try to normalize data by using the logarithm or other transform of the variable. Normal, or Gaussian statistics are known as parametric statistics. see **Non-parametric statistics**

NORTH ATLANTIC OSCILLATION
High pressure over the Azores and lower pressure over Iceland affect weather in Europe. A steeper pressure gradient pushes the jet stream further north and the weather over western Europe tends to be milder. see **El Nino/Southern Oscillation**

NORTHEAST STATES FOR COORDINATED AIR USE MANAGEMENT (NESCAUM) (US)
A cooperative group set up among eight states for dealing with air pollution problems common to member states. Policy statements and standards are issued periodically.

"NOT IN MY BACKYARD" (NIMBY)

A term applied to opposition to local siting of desirable or necessary, but possibly not popular, projects such as incinerators.

NO$_x$

see **Nitrogen oxides**

NPDES

see **National Pollutant Elimination System**

NPL

see **National Priority List**

NRA

see **National Rivers Authority**

NRC

see **Atomic Energy Act**

NRD

see **Natural resource damages**

NRPB

see **National Radiation Protection Board**

NSPS

see **New Source Performance Standards**

NTA

see **Nitrilotriacetic acid**

NUCLEAR REGULATORY COMMISSION

see **Atomic Energy Act**

NUCLEAR SPECIES

see **Nuclide**

NUCLEAR WASTE POLICY ACT (US)

Passed in 1982 and reauthorized in 1987, this Act directs the Department of Energy (DOE) to find a suitable site for and design and construct a depository for high level radioactive waste and spent fuel from civilian nuclear reactors.

NUCLEIC ACIDS

The major function of these is to store, maintain and transmit genetic information. DNA and RNA are important nucleic acids.

NUCLIDE

A species of atom defined by the energy content and the number of protons and neutrons.

NUTRIENT CYCLE
see **Biogeochemical cycle**

NUTRIENTS

All organisms require nutrients for growth. The major nutrients of environmental significance are carbon, hydrogen, oxygen, nitrogen and sulfur. Nitrogen and phosphorous compounds are of interest because, at elevated concentrations, they can cause health and eutrophication problems.

OBLIGATE PHOTOTROPHS

Organisms which are completely dependent upon photosynthesis. They cannot use organic compounds for growth in the dark.

OBLIGATE POIKILOTHERM

Thermal conformer. An organism which is able to exert little significant control of body temperature by specialized metabolism or behavior.

OCCUPATIONAL HEALTH

This discipline is concerned with the relationships between work environment and practices and health. A joint ILO/WHO committee has defined occupational health as the promotion and maintenance of the highest degree of physical, mental and social well-being of workers in all occupations. Occupational (industrial) hygiene is the application of scientific principles to improving a working environment.

OCCUPATIONAL HEALTH AND SAFETY ACT (OSHA) (US)

Employers are required to provide each employee a workplace free of recognized hazards which are likely to cause illness, injury or death. The Department of Labor can promulgate standards on safety and health, conduct inspections, issue citations and impose penalties. Employers are required to maintain records.

OCEAN DUMPING ACT

see **Marine Protection, Research and Sanctuaries Act**

OCEANOGRAPHY
The study of the origins of and physical, chemical and biological characteristics of the world's oceans.

OCEAN THERMAL ENERGY CONVERSION
Temperature differences of 20°C between cooler deeper ocean water and warmer solar energy absorbing surface water allow development of a thermal source (engine) which can be used to produce other forms of energy. Major disadvantages are cost and scale.

OFFICIAL METHODS OF ANALYSIS OF THE ASSOCIATION OF OFFICIAL AGRICULTURAL CHEMISTS
As with Standard Methods, this is a collection of analytical methods approved by a board of experts. It is necessary to have widely accepted methods so that results are consistent and credible. see **Standard Methods**

OIL POLLUTION ACT OF 1990 (US)
This Act mandates that all tank vessels of 5000 gross tons (4536 metric tonnes) or larger operating in waters subject to jurisdiction of the United States must have double hull construction by 2015. Phase-out schedules for existing vessels, depending on year of delivery and cargo tank configuration, are included.

OLFACTORY
Dealing with the sense of smell.

OLIGOTROPHIC
A state of low productivity.

OPACITY
The degree to which light transmission is reduced or visibility of a background is reduced when viewed through a plume.

OPEN CHANNEL FLOW
The free surface of the water is open to the atmosphere. The flow is not under pressure.

OPERATIONAL MISHAPS
Spills into surrounding waters occurring during pumping of fuel oil bunkers and liquid cargo.

OPTICAL DENSITY

In a spectrophotometer, the percentage of light transmitted through a sample is compared to the light transmitted through a blank sample containing reagents of the test in question. Light through the blank is adjusted to read 100% transmittance. Use of optical density (absorbance), defined as log(100%/% transmittance), allows arithmetic plotting of both scales of a calibration curve. If Beer's Law is followed, a plot of arithmetic concentration versus arithmetic optical density (O. D.) gives a linear response. Arithmetic concentration plotted against the log of % transmittance also gives a straight line. see **Beer's Law**. see **Lambert's Law**. see **Spectrophotometer**

ORGANIC CARCINOGENS

Carbon compounds which cause cancer in experimental animals and are, therefore, suspected of playing a role in human cancer.

ORGANIC COMPOUNDS

These are compounds which contain carbon. It was first thought that organic compounds could only be produced by plants or animals. The great number of organic compounds results from the valence of four of the carbon atom. Carbon atoms can link together by covalent bonding. Organic compounds are usually combustible and reactions are molecular rather than ionic. Several compounds can exist with the same formula (isomerism). Most organic compounds can serve as food for microorganisms.

ORGANOCHLORINE

Bleaching of paper stock by pulp mills produces organochlorines. More than 350 organochlorines have been found in pulp mill effluents. Known as AOX, these substances have been shown to have adverse effects on aquatic life. The abatement approach is control of the discharge collectively rather than dealing with individual components.

ORGANOCHLORINE INSECTICIDES

Included in this group are DDT, BHC, toxaphene, aldrin, dieldrin, endrin, heptachlor and chlordane. Dominating the market for more than 25 years, these were considered to be too damaging environmentally. In the United States legal restrictions on DDT were introduced in 1973 and by 1984 use of organochlorine insecticides was phased out almost completely. see **Dichlorodiphenyltrichloroethane**

ORGANOPHOSPHOROUS

The most important pesticides among these compounds are malathion and parathion. They are an improvement over organochlorine compounds because they decompose fairly rapidly and do not accumulate in the food chain. Parathion is quite toxic to humans while malathion is about 100 times less toxic.

ORGANOTIN

These compounds found wide use in antifouling paints. Tributyl tin is considered very toxic to shellfish and can accumulate in bottom sediments and up the food chain. These are no longer permitted for application to small boats. They have been replaced by copper oxides with added biocides. However, the copper and biocides are of concern.

ORP

see **Oxidation–reduction potential**

OSHA

see **Occupational Safety and Health Act**

OSHA CRITERION LEVEL. (US)

90 dBA for an 8 hour day. The permissible exposure limit is halved for each 5 dBA sound level increase.

OSLO CONVENTION

see **Convention for Protection of Marine Resources of the North-East Atlantic**

OSMOSIS

Movement of a solvent through a membrane which is impermeable to a solute. Reverse osmosis is used for production of potable water and treatment of some wastewaters.

OSPAR

see **Convention for Protection of Marine Environment of the North-East Atlantic**

OSPAR COMMISSION

Work under the OSPAR Convention is managed by representatives of governments of the contracting parties and the European Commission.

The ecosystem approach is applied for management of human activities. The six topics addressed are protection and conservation of marine biodiversity and ecosystems; eutrophication; hazardous substances; offshore oil and gas; radioactive substances; and monitoring and assessment.

OTEC
see **Ocean thermal energy conversion**

OUTER CONTINENTAL SHELF LANDS ACT
see **General Mining Act**

OVERTURNS
Water is densest at 4°C. In summer and winter freshwater bodies are often thermally stratified and quite stable. In the spring and fall this thermal stratification is upset and vertical circulation is induced, primarily by wind action. There may be spatial water quality variation in a reservoir and it is often desirable to draw water from different depths.

OXIDATION PONDS
Sometimes called sewage lagoons, these are simply large shallow ponds to which sewage wastes are added at a single point. Usual depths are 2 to 4 feet (0.6–1.2 m), the minimum depth being controlled by weed growth. Mixing is important and aerators provide greater agitation than do wind generated currents. Oxidation ponds have found wide application in Europe but have not been well accepted in North America, primarily on esthetic grounds.

OXIDATION-REDUCTION POTENTIAL
The voltage difference generated in a reversible oxidation-reduction system. This is an important parameter in operation of the activated sludge process.

OXIDIZING AGENT
Oxidation involves a gain in electrons. Oxidizing agents are those which give up oxygen or combine with a reducing agent. Chlorine is a strong oxidizer.

OXISOLS
see **Tropical soils**

OXYGENATED HYDROCARBONS

Included among these compounds are alcohols, ethers, phenols, aldehydes, ketones, peroxides and organic acids. These are quite reactive and readily form photochemical oxidants and oxides of nitrogen in the presence of sunlight. Many are themselves toxic and some, particularly esters and ketones, can cause central nervous system disorders. Many are known carcinogens.

OXYGENATES

Substances added to gasoline to increase the octane and give lower carbon monoxide and unburned hydrocarbons in the exhaust. Oxygenates increase combustion temperatures. The most widely used oxygenate is methyl tert-butyl ether (MTBE). When fuel containing oxygenates enters the groundwater from leaking underground storage tanks the resulting groundwater pollution is difficult to treat and this has caused manufacturers and distributors problems.

OXYHEMOGLOBIN

see **Methemoglobinemia**

OXYGEN SAG (STREETER-PHELPS) EQUATION

When waste is introduced into a receiving water, two simultaneous actions are set up. Exertion of BOD takes oxygen from solution. At the same time, the lowered oxygen concentration sets up a driving force trying to bring oxygen back into solution. The DO deficit, D, is the difference between the saturation concentration and the actual concentration. Streeter and Phelps proposed the fundamental expression still in use. Terms added later dealt with algal contribution to dissolved oxygen and benthal (bottom) demand.

OZONATION

Application of ozone to stabilization of wastewater and wastewater sludge. This is a promising approach but has not been used widely.

OZONE

An unstable form of oxygen, O_3, ozone is an important part of photochemical smog. It originates primarily from the effect of sunlight on automobile exhaust gases. 90% of the planet's ozone exists in the stratosphere. Ozone absorbs all of the solar radiation between 240 and 290 nm and this radiation is lethal to one-cell organisms and surface cells of higher plants and animals. In addition, it reduces solar radiation up to 320 nm. These wavelengths are biologically active and reduction of stratospheric ozone may lead to skin cancers in susceptible individuals due to greater ultraviolet exposure. Stratospheric ozone is the major heat source for the stratosphere due to absorption of ultraviolet, visible and infrared radiation. It is thought that changes in stratospheric ozone may lead to significant climate changes. The "ozone hole" over the South Pole is causing great concern and there is much controversy as to the extent of damage to the ozone layer. Ozone has found some use in stabilization and volume reduction of wastewater sludge.

OZONE HOLE

Over Antarctica the ozone column was found to have decreased markedly from the 1950s to the 1990s. The bulk of the effect on ozone abundance is found in the lower stratosphere, between 15 km and 22 km. Ozone decreases during the winter and returns closely to normal with springtime warming. There is controversy as to whether the ozone hole is spreading and no completely adequate explanation of this phenomenon has been advanced. However, the first Nobel Prize awarded on a topic dealing with the environment was awarded for work on the chemistry of ozone destruction in the stratosphere.

OZONE ISOPLETHS

A contour diagram used to illustrate the way in which the maximum ozone concentration reached over a fixed irradiation period depends on initial NO_x and reactive organic concentrations. These isopleths are useful in evaluating control strategies.

P

PACKAGE PLANTS
Complete wastewater treatment units manufactured for small communities and installations not served by municipal systems. These units, which do not require continuous operational supervision, supply primary or some biological treatment. see **Primary treatment**. see **Secondary treatment**

PACKED BED
Solid particles placed in a column in either orderly or random fashion to give a large surface area for mass transfer. See **Fluidized bed**. see **Particulate scrubber**

PACKED TOWER
see **Packed bed**

PAH
see **Polynuclear aromatic hydrocarbons**

PAN
see **Peroxyacetyl nitrate**

PARAMETRIC STATISTICS
see **Normal distribution**

PARAMETRIC VALUES
Maximum permitted concentrations, usually in micrograms per cubic meter, for chemicals that are toxic or carcinogenic.

PARASITISM
Microorganisms cause disease in the host by taking necessary nutrients from the host organisms.

PARATHION
see **Organophosphorous**

PARIS CONVENTION
see **OSPAR Convention**

PARSHALL FLUME
An open channel device for measuring flow into a wastewater treatment plant, it is also called an open channel venturi. see **Critical flow**

PARTICULATES
Dispersed solid or liquid matter larger than single molecules (0.0002 microns) but smaller than 5000 microns. Particulates in the atmosphere range in diameter from 0.1 micron to 10 microns. Particulates are made up of products of incomplete combustion, metals, large ions, mists, dusts and acid mists. While particulates cause soiling of surfaces and affect visibility, the greatest effects are on human health. Small particulates can cause lung irritation and reduce lung respiratory efficiency by inhibiting transport of oxygen from the lungs. Particulates of health interest are commonly written as P_{10}, $P_{2.5}$ and $P_{0.1}$. These are particles which can be taken into the lungs. The smaller the particle, the deeper it can travel in the lungs. Particulates can carry adsorbed toxic materials on their surface.

PARTICULATE SCRUBBER
A wet collection device for removal of fumes, mists and suspended dusts.

PASCAL
Pressure resulting from applying a force of 1 newton uniformly over an area of one square meter.

PASQUILL-GIFFORD GRAPHS

A scheme for estimating vertical and lateral air pollution plume widths as functions of downwind distance and meteórological conditions. These are now used less than they used to be.

PASTEURIZATION

Raw milk can transmit a number of diseases. This process, developed by Louis Pasteur, does not sterilize but can disinfect milk. Any disinfection process must not change significantly the chemical nature of the milk. It was found that milk brought to 143°F (61.6°C) for 30 minutes or 161°F (71.7°C) for 15 seconds remains palatable while pathogens are destroyed. see **Phosphatase test**

PATHOGENS

Microorganisms which are capable of producing disease in humans, plants and animals are known as pathogenic. The disease may occur through metabolic upset of normal metabolic reactions in the host. This can be through parasitism, where the pathogen derives nutrients at the expense of the host organism, or by production of toxins. Common pathogens found in water include *Salmonella typhosa* (*E. typhosa*), *Shigella dysenteriae* and *Vibrio comma.* *Entamoeba histolytica* is a pathogenic protozoan.

PCBS

see **Polychlorinated biphenyls**

PCV

see **Positive crankcase ventilation**

PEARL CHAIN EFFECT

Particles align themselves in a chain when subjected to an electric field. This phenomenon is of interest in the study of biological effects of microwave radiation. see **Microwave radiation**

PEL

see **Permissible exposure limits**

PERCOLATING FILTER

see **Trickling filter**

PERCOLATION
Water moving downward by gravity through soil.

PERCOLATION TEST
This is a test to determine suitability of soil to accept effluent from a tile field.

PERC TEST
see **Percolation test**

PERMEABILITY
A measure of the ability of a granular material to allow water flow.

PERMISSIBLE EXPOSURE LIMITS (PEL) (US)
These are set under the Occupational Safety Act. Limits are set for protection of workers against exposure to hazardous workplace substances and conditions. PELs must be both technically and economically feasible.

PEROXYACETYL NITRATE (PAN)
A photochemical oxidant responsible for much of the plant damage associated with photochemical oxidants. It is an eye irritant. PAN is relatively stable in the upper troposphere and acts a reservoir for nitrogen oxides.

PERVAPORATION
A contraction of permeation and evaporation, this is a method for removal of volatile organic carbons (VOCs) from wastewaters. A selective polymer absorbs the contaminant, which is then desorbed by evaporation on the permeate side of a preferential absorption membrane. Continuous absorption on one side and desorption on the other creates a mass flow rate through the membrane.

PESTICIDE
An economic poison used to control or destroy pests which cause economic loss or adverse human health effects. Pesticides include insecticides, herbicides, fungicides, iacaricides, nematocides, repellants, attractants and plant growth regulators. Some are metabolized slowly and can accumulate in adipose tissue.

PET

see **Polyethylene terephthalate**

pH

A shorthand expression for hydrogen ion concentration. pH is calculated as the negative logarithm of the hydrogen concentration of the solution. Pure water has a pH of 7. Values less than 7 are acidic and values above 7 are alkaline. $pH = log_{10}1/(H^+) = -log_{10}(H^+)$.

PHAGOTROPHS

Heterotrophic organisms which ingest other living organisms or organic particulate matter.

PHENOL ($C_6 H_5OH$)

Hydroxy derivative of benzene. Phenol in water is not discerned at concentrations of the order of 1 mg/L. However, chlorination of water with a phenol concentration will give an objectionable taste due to formation of chlorophenol.

PHORETIC PHENOMENA

Particle motion which occurs when there is a difference in the number of molecular collisions on different sides of the particle.

PHOSPHATASE TEST

Adequacy of milk pasteurization is controlled by determining the presence of the enzyme phosphatase. It is heat sensitive in normal pasteurization temperature ranges. Addition of CQC causes a blue color if pasteurization is not complete and all phosphatase has not been destroyed. see **Pasteurization**

PHOSPHATES

Formerly, large amounts of phosphates were used as builders in detergents. These water softening compounds are contributors to eutrophication and are no longer used in detergents. Organisms responsible for wastewater treatment require phosphorous for reproduction and cell synthesis. Phosphates are used in public water supplies for controlling corrosion.

PHOSPHOROUS

A solid non–metallic element existing in the poisonous and flammable yellow form and the red form. The red form is less poisonous and less flammable. Yellow phosphorous is a protoplasmic poison.

PHOTOCHEMICAL OXIDANTS

Secondary pollutants which result from a complex series of atmospheric reactions involving organic pollutants, NO_x, oxygen and sunlight. Main photochemical oxidants are ozone, NO_2 and peroxyacetyl nitrate.

PHOTIC ZONE

see **Euphotic zone**

PHOTOSYNTHESIS

Use by green and purple pigmented microorganisms (autotrophs) of light and CO_2 as sources of all or part of their energy. Electrons are split from chlorophyll and are attached to a coenzyme, along with a hydrogen ion. The resulting hydrogen ion serves as fuel for the cell.

PHOTOVOLTAIC CELLS

Electrical current sources driven by a flux of radiation. Also called solar cells, these devices find use in special applications. See **Renewable energy**

PHREATIC WATER

Groundwater.

PIT OR VAULT PRIVY

see **Human waste**

PLANKTON

Drifting and free floating algae and animals found in lakes, oceans and streams. Some organisms become colonial when numbers increase. A "bloom" occurs as a result of a rapid increase in numbers of plant cells. One definition of bloom is 500 individual cells per ml.

PLASMODIUM VIVAX

The causative protozoan organism of malaria. Malaria is caused by the bite of an infected female anopheles mosquito.

PLASTICS

Polymeric materials. When discarded, these constitute a significant part of the solid waste stream and are suitable for recycling. Plastics are labeled for recycling.

1   PET – Polyethylene terephthalate
2   HDPE – High density polyethylene
3   V – Vinyl
4   LDPE – Low density polyethylene
5   PP – Polypropylene
6   P – Polystyrene
7   Other

Some municipal recycling centers will accept only 1 and 2.

PLASTICS TO DIESEL
Mixed plastics waste is pyrolized. The pyrolysis gas is then condensed into a hydrocarbon distillate composed of straight and branched chain aliphatics, cyclic aliphatics and aromatics. Sulfur content is low.

PLATFORM ABANDONMENT
The International Maritime Organization has issued documentation entitled "Guidelines and standards for the removal of offshore installations and structures on the Continental Shelf and the exclusive economic zone" dealing with removal of abandoned or unused drilling platforms. These deal with navigational safety and environmental protection.

PLUG FLOW
No element of flowing fluid overtakes another element. This is also known as non-mix flow.

PLUMBOSOLVENT
see **Lead**

PMN
see **Premanufacture Notice**

PNEUMATIC BOOM
Also called a bubble barrier, this is a permanently installed means of dealing with potential oil spills. A perforated pipe through which air can be pumped is located below the surface in areas where petroleum products are handled. Bubbles rising to the surface form a barrier against spread of contaminant. This cannot be used where there are strong currents.

PNEUMOCONIOSIS

Known as black lung disease, this affects the lungs of workers exposed over long periods to mineral dust. These dusts include silica, coal and asbestos.

POINT SOURCE

A conveyance by which pollution reaches a receiving water.

POLDER

Land below sea level which is reclaimed for agricultural and other uses.

POLLUTANT

A substance which has an adverse effect on the environment.

POLLUTER PAYS PRINCIPLE (UK)

This originated in Yorkshire early in the twentieth century as a result of an Act of Parliament. Now the cost to the producer of an industrial waste is determined by the Mogden formula. see **Mogden formula**

POLLUTION

Adverse effects upon the environment. According to the National Research Council (US), "Pollution is a resource out of place."

POLYCHLORINATED BIPHENYLS (PCBs)

These compounds are man-made and do not exist in Nature. The biphenyl nucleus contains 10 replaceable hydrogen atoms and there are 209 isomers in 10 homologous series. PCBs are heat stable, have no flash point, are very stable and unreactive, only slightly soluble in water, very soluble in oils and organic solvents and have low vapor pressures. Commercial PCBs have been produced worldwide under a variety of trade names. A common use was in electrical transformers. Based on the large amount of data published in the scientific literature, the adverse health effects attributed to PCBs are probably due to the presence of trace amounts of polychlorinated dibenzofurans (PCDFs). PCBs are members of the larger group of askarels.

POLYCHLORINATED DIBENZOFURANS (PCDF)

In the manufacturing process for PCBs there can be contamination with polychlorinated dibenzofurans. If the contaminated PCBs are introduced into food, as happened in the Yu-Cheng incident and

Yusho disease, there can be serious effects. Based on available scientific reports, these health effects are due to the PCDF impurity. see **Yu-Cheng incident**. see **Yusho disease**

POLYCYCLIC AROMATIC COMPOUNDS
see **Polynuclear aromatic hydrocarbons**

POLYELECTROLYTES
Long chain molecules used as aids in flocculation.

POLYETHYLENE TEREPHTHALATE (PET)
A common plastic used for beverage containers. When recycled, these can be used for containers which will not be in contact with food.

POLYMERS
These are formed, in a series of steps, by linking smaller molecules called monomers.

POLYNUCLEAR AROMATIC HYDROCARBONS (PAH)
These evolve from high temperature reactions under pyrolitic conditions during incomplete combustion. Included are benzopyrene, benzo-anthracene, benzoacetophenthrylene, chrysene and benzo-(a)-pyrene.

POLYSACCHARIDE
Monosaccharides (tetrose, pentose and hexose) can polymerize to form polysaccharides. Polysaccharide molecules form the capsule protecting the microbial cell. see **Green–Stumpf theory**

POROSITY
The ratio of the volume of voids to the total volume.

POSITIVE CRANKCASE VENTILATION (PCV)
A means of reducing automobile emissions by pulling fumes and air into the combustion chamber by means of vacuum.

POST-CONSUMER RECYCLING
Recovery of materials from municipal solid waste streams. Most commonly recycled are newspapers, glass, aluminum cans, corrugated cardboard, ferrous and non-ferrous metals and plastics. These materials are collected at the curbside or are carried to a central collection point.

POTABLE WATER

Water intended for drinking and other high quality uses such as cooking.

POTENTIATION

The process by which one compound is made more toxic due to the presence of a second compound. The opposite of this response is antagonism. see **Antagonism**

POTW

see **Publicly owned treatment works**

PPP

see **Polluter Pays Principle**

PRECIPITATION

Aqueous material reaching the ground as rain or snow. Frost and fog are included in this category but are not measured routinely.

PREMANUFACTURE NOTICE (PMN) (US)

Under provisions of TSCA, a person who proposes to import, manufacture or process a new chemical or introduce a significant new use of an existing chemical must notify the Environmental Protection Agency.

PREVENTION OF SIGNIFICANT DETERIORATION (PSD) (US)

Under the Clean Air Act, new sources of pollution are regulated in areas where standards are presently met.

PRIMARY IRRITANTS

Agents which induce local, minor to severe responses and, at times, death of cells.

PRIMARY POLLUTANTS

Those which are discharged directly from identifiable sources.

PRIMARY PRODUCERS

Organisms which are autotrophic. These organisms are able to synthesize their own food from simple inorganic substances, using chlorophyll and energy from sunlight or from chemical oxidation of inorganic compounds. Total energy or matter in metabolism is gross primary production and that incorporated as new tissue is net primary production.

PRIMARY PRODUCTION
Chemical energy produced by green plants per unit area per unit time.

PRIMARY TREATMENT
Physical and mechanical treatment of wastewater, including screening, sedimentation and oil and grease removal. BOD removal is about 30%. Chemical treatment alone is considered primary. see **Secondary treatment**. see **Tertiary treatment**

PRIORITY POLLUTANTS
In the 1970s the US EPA created a list of 129 pollutants of particular environmental significance. Criteria for inclusion were toxicity, how common was the chemical, likelihood of concentration and movement through the environment. In the UK a comparable List would be the Red List. see **Emerging contaminants. See Red List**

PROCESS LOADING FACTOR
see **M/F ratio**

PROCESS PERFORMANCE
The percentage ratio between the substrate (food) removed and the influent substrate concentration.

PROHIBITED DISCHARGES
Substances which can cause fire or explosion hazard, obstruction or corrosion are prohibited from discharge to sewers and Publicly Owned Treatment Works (POTW). see **Publicly Owned Treatment Works.** see **Sewer ordinance**

PROKARYOTES
These do not have genetic material separated from the rest of the cell by a membrane. Included are viruses, mycoplasma, true bacteria, spirochetes, actinomycetes, mycobacteria, budding bacteria, gliding bacteria and blue-green algae.

PROTEINS
Complex compounds of carbon, hydrogen, oxygen and nitrogen made up of amino acids joined by peptide linkages. Sulfur and phosphorous

may also be constituents. Proteins are necessary dietary components for higher animals. Molecules are large and treatment of wastes containing proteins will need to be biological in nature.

PROTISTA

Single-celled organisms which have both plant and animal characteristics. They range from viruses (30–300 nm) to protozoa (500–50 000 nm). They are divided into prokaryotes, which do not have genetic material divided from the rest of the cell in a membrane, and eukaryotes, which have a true nucleus.

PROTOZOA

Single celled animals which reproduce by binary fission. The majority metabolize solid organic food and are aerobic. These organisms are classified in five groups. (1) *Sarcodina*. Included is the pathogen *Entamoeba histolytica*. (2) *Mastigophora*. These move by means of flagella. (3) *Sporoza*. These are parasitic and include *Plasmodium vivax*, which causes malaria. (4) *Ciliata*. These move by hairlike cilia and are the most important protozoan with respect to stream pollution and waste treatment. *Paramecium* is a member of this group. (5) *Suctoria*. Phases of the life cycle are a ciliated free swimming early stage and in the adult a stalked stage.

PSD
see **Prevention of significant deterioration**

PSYCHODA FLY

A small insect which breeds in trickling filter slime. The fly does not bite but can get into the nose and ears and is quite bothersome. Control is by periodic flooding of the filter or chlorination of influent.

PUBLIC HEALTH

The art and science of preventing disease, prolonging life and promoting physical and mental health.

PUBLIC HEALTH SERVICE DRINKING WATER STANDARDS (US)

These were promulgated in 1962 and set forth levels for contaminants in potable water. The highest use is set as human consumption.

PUBLICLY OWNED TREATMENT WORKS (POTW) (US)

Wastewater treatment facilities owned and operated by agencies at all levels of government. Regulations under the National Pollutant

Discharge Elimination System (NPDES) apply to these plants. Discharges prohibited to sewers serving these plants are those which can cause fire or explosion, corrosion, slug discharges and heat discharges. see **Prohibited discharges**

PULMONARY
The portion of the respiratory tract which includes the respiratory bronchiole, alveoli ducts and alveoli.

PUMP
A device for moving fluid from one point to another. External energy is converted to kinetic and potential energy. see **Total dynamic head**

PUMP CURVE
A plot of volumetric discharge versus head loss.

PUMPED STORAGE
Electric utilities have considerable fluctuation in power demand, depending on time of day. In order to avoid provision of costly additional generating capacity to meet peak demand, some utilities use excess power during periods of low demand to pump water to storage reservoirs. This water then becomes available for use in hydropower generation during periods of high electric usage.

PURE TONE
Sound consisting essentially of a single-frequency sinusoidal pressure wave.

PYRITE
The iron sulfide mineral primarily responsible for acid mine drainage. Upon oxidation, the FeS_2 forms sulfates and sulfuric acid. The oxidation paths are complex and not completely understood.

PYROLYSIS
Heating of waste in a closed vessel in the absence of air. The process is complicated and requires gas purification, making it impractical on a small scale.

Q

QAC

see **Quaternary ammonium compounds**

Q_{10}

The ratio of the reaction rate at a particular temperature with respect to the rate at a temperature 10°C lower.

QUATERNARY AMMONIUM COMPOUNDS

Widely used disinfecting agents which have strong germicidal properties. Cationic detergent usually indicates the presence of a QAC. QACs are organically substituted ammonium compounds which function best under alkaline conditions.

R

RACKS

These are provided at the head of a wastewater treatment plant to remove large floating objects such as wood and rags.

Rad (ROENTGEN-ABSORPTION-DOSE)

Unit of ionizing radiation corresponding to energy absorption of 100 ergs (1×10^{-5} J/s) per gram. The rad is a massive radiation dose to a person and another unit for exposure of humans to radiation has been developed. This is the rem (roentgen-equivalent-man) and corresponds to the amount of radiation that produces an energy dissipation in the human body that is equivalent biologically to one roentgen of X-rays. The rad has been replaced by the Gray (Gy), which is equal to 100 rads.

RADIATION CONTROL FOR HEALTH AND SAFETY ACT (US)

The declared purpose of this Act is the establishment of a national electronic product radiation control program. It covers X-rays, gamma, ultraviolet, visible, infrared, radio frequencies and microwaves. Performance standards have been promulgated for TV sets, microwave ovens and lasers.

RADIATION ECOLOGY

see **Radioecology**

RADIOACTIVE SUBSTANCES ACT (RSA) (UK)

This comprehensive legislation dealing with radioactive waste was passed by Parliament in 1960 and implemented in 1963. It was amended by the Environmental Protection Act of 1990 and further amended in 1993.

RADIOACTIVE WASTE

A solid, liquid or gaseous material of negligible economic value containing radionuclides in excess of threshold values. High level waste (HLW), produced in the first cycle of reprocessing spent nuclear fuel is highly radioactive. Intermediate level wastes (ILW) can be divided into short-lived, with half lives of 20 years or less, and long-lived, whose principal constituents have half lives of thousands of years or more. Low level wastes (LLW) contain less than 4 GBq/ton (0.9×10^{12} Bq/kg) of alpha emitting radionuclides and less than 3.6×10^{12} Bq/kg of beta and gamma emitters. Very low level wastes (VLLW) contain activity concentrations less than 0.4 MBq/ton.

RADIOACTIVITY

Spontaneous disintegration of an element. The heaviest elements emit radiation of short wavelength (gamma rays), fast electrons (beta rays) and helium particles (alpha rays).

RADIOECOLOGY

That area of the broad field of ecology concerned with assessment of radiation and radioactive substances and the environment.

RADIONUCLIDE

Radioactive isotope.

RADON

An odorless and colorless naturally occurring radioactive gas which can seep into dwellings through basements. One of the products of the decay chain from uranium-238, it is normally reported as picocuries per liter or becquerels per cubic meter.

RAINFALL INTENSITY CURVES

These plots relate rainfall intensity (in/hr, mm/hr) and storm duration. From records for the area under consideration, return periods for various intensities are derived and displayed as separate curves on the graph.

RAIN FOREST

Dense forests growing in tropical areas with high rainfall. In climatological terms, a rain forest is defined as having continuously high temperatures, high annual rainfall and diurnal temperature variation often higher than the annual variation. 12 million hectares are lost each year to logging and cut and burn for agricultural land. These losses are considered very serious because of the effect on climate and destruction of species, with attendant loss of genetic blueprints.

RAINOUT
Scavenging of chemicals from in-cloud water.

RAMSAR LIST
see **Convention on Wetlands**

RAMSDELL EQUATION
An expression relating collector efficiency and number of active bus sections in an electrostatic precipitator.

RANDOMIZATION
A sample for statistical analysis taken in such a way that uncontrolled variables which might affect results have an equal chance of affecting any of the variables.

RAPID SAND FILTER
Water flows through a stratified sand bed 12 to 30 inches (30–75 cm) in depth. The bed is supported by a 6 to 18 inch (15–45 cm) layer of gravel or other coarse material. Removal of floc from the flowing water is due primarily to adsorption onto the sand grains and straining is not the principal mechanism. The sand bed is washed when head loss becomes excessive due to clogging of the upper layer or dirty water is short circuited. The bed is cleaned by backwashing with clean water and it is not necessary to take the unit out of service, as is the case with a slow sand filter. Clean water is pumped upward through the filter from the bottom and the bed is suspended in the flowing water, forming a fluidized bed. Adsorbed floc is carried to waste. The bed settles into stratified layers when the washwater flow ceases. The pores at the top are smallest and clog most easily. Use of several materials of different densities and porosities allows longer filter runs than are possible using sand alone.
see **Fluidized bed**. see **Mixed media filter**. see **Slow sand filter**

RAR
see **Reasonable assumed resource**

RAYLEIGH SCATTERING

The phenomenon of light scattering by aerosol particles. This has strong size and wavelength dependence, with greatest scattering at short wavelengths.

RBC

see **Rotating biological contactor**

RCRA

see **Resource Conservation and Recovery Act**

RDF

see **Refuse-derived fuel**

REASONABLE ASSUMED RESOURCES (RAR)

Deposits of sufficient size and configuration that recovery is economically feasible, using current technology.

RECALCITRANT COMPOUND

One which is not easily decomposed by microorganisms.

RECIRCULATING TOILETS

One means of dealing with the prohibition against overboard discharge of untreated waste from watercraft is to use a closed system charged with a disinfecting substance. Deposited waste is stored in a retention tank. For subsequent uses, a portion of the disinfectant/waste mixture is used for flushing. This technology has been used for many years aboard commercial airliners. After 80–100 uses, the retention tank must be emptied into a shore based pump-out station.

RECIRCULATION

That portion of biologically treated wastewater which is sent back to the head of the plant. Recirculated water contains microorganisms necessary for keeping the process active.

RECYCLING

Collection and treatment of a waste product for reuse as a raw material in manufacturing of the same or different product.

RED DATA BOOKS

see **International Union for Conservation of Nature and Natural Resources**

RED LIST (UK)

This is a list of the most dangerous substances as established by scientific evidence and criteria. On the Red List are aldrin, atrazine, azinphos-methyl, cadmium and cadmium compounds, DDT, DDD, DDE, 1,2-dichloroethane, dichlorvos, dieldrin, endosulfan, endrin, fenitrothion, hexachlorobenzene, hexachlorobutadiene, lindane, malathion, mercury and mercury compounds, polychlorinated biphenyls (PCBs), pentachlorophenol, simazine, trichlorobenzene, triflualin and triorganotin compounds. see **Black List**. see **Grey List**

REDOX

see **Oxidation–reduction potential**

RED TIDE

Reddish discoloration of coastal surface waters due to toxin-producing dinoflagellates.

REDUCING AGENT

Reduction is loss of electrons and reducing agents are those which remove oxygen from a compound or which add hydrogen.

REFORMULATED GASOLINE (RFG) (US)

Motor fuels to which oxygenates have been added to raise octane level, combustion temperatures and engine efficiencies. According to the Clean Air Act Amendment (CAAA), these fuels must contain 2% by weight of oxygen, no more than 1% benzene and 25% total aromatics. Areas of the US which fail to meet national ozone standards will be required to use RFG the year round. Additives include MTBE (methyl tert-butyl ether), ETBE (ethyl tert-butyl ether), TAME (tert-amyl methyl ether) and TAEE (tert-amyl ethyl ether).

REFRACTORY ORGANIC

A man-made organic compound, such as a pesticide, that degrades slowly.

REFUSE

All solid waste matter.

REFUSE-DERIVED FUEL (RDF)

Raw refuse is separated into organic, paper, and recoverable and recyclable (glass, metals, plastic, etc.) portions. Combustible material is shredded and can be used in this lowest form of RDF. The material can be pelletized and co-fired with coal. Pelletizing is an extra step but does reduce handling and storage problems.

RELATIVE HUMIDITY
The ratio (expressed as a percentage) of the amount of water vapor present in the air to that present at saturation.

Rem (ROENTGEN-EQUIVALENT-MAN)
The amount of radiation which produces energy dissipation in the human body equivalent biologically to one roentgen of X-rays. The rem has been replaced by the Sievert (Sv), which is equal to 100 rem.

REMEDIAL INVESTIGATION (US)
The process under CERCLA for determining the extent of hazardous waste contamination. This may include treatability studies.

RENEWABLE ENERGY
Fossil fuels are finite. Some forms of energy are available and will not be exhausted by one-time use. Among these are solar energy, photo-voltaic generation, hydropower, wind power, biofuels, wave energy, tidal power, geothermal and nuclear energy.

RENEWABLE TRANSPORT FUELS OBLIGATION (UK)
New measures require that 5% of motor fuel sold in the UK must be from a renewable source by 2010.

Rep (ROENTGEN-EQUIVALENT-PHYSICAL)
The quantity of radiation, other than gamma or X-radiation, which produces in one gram of human tissue ionization equivalent to that produced in one roentgen of X or gamma radiation.

RESERVOIR ROUTING
see **Flood routing**

RESOURCE CONSERVATION AND RECOVERY ACT (RCRA) (US)
This Act regulates and governs management of hazardous waste in the US. Enacted in 1976, it was amended in 1984. The approach involves federal identification of hazardous waste from generator to ultimate disposal, setting minimum standards for treatment and storage and state

implementation of hazardous waste plans at least as stringent as in the federal program. Included are provisions for municipal recycling programs.

RESPIRABLE FIBER
A fiber below 3 microns in diameter and longer than 5 microns.

RESPIRABLE PARTICLE
Airborne particles which are most likely to be taken into the lung. Particles below 7 microns will reach the alveoli of the lung.

RESPIRATION
A measure of metabolic activity.

RESPIRATORY TRACT
Included are three tracts–nasopharnyx, tracheobronchial and pulmonary.

RETENTION BASIN
see **Detention basin**

RETURN PERIOD
The probability that an event will occur in a given period. A 100 year return period means that there is a 1% probability of occurrence in any one year.

REVERSE OSMOSIS
Water is forced to flow through a semi-permeable membrane from a region of higher salt concentration by pressure higher than osmotic pressure. Reverse osmosis is used in desalination of brackish and fully saline waters. see **Desalination**

REYNOLDS NUMBER (Re)
A dimensionless number which describes the flow regime under consideration and allows comparison of flow among different fluids. For pipes, a Reynolds number less than 2000 describes laminar flow and turbulent flow exists at Reynolds numbers greater than 4000. The transition from laminar to turbulent flow occurs in the Re range of 2000 to 4000. see **Laminar flow**. see **Turbulent flow**

RFG
see **Reformulated gasoline**

RHIZOSPHERE
Soil that surrounds and is influenced by the roots of a plant.

RI
see **Remedial investigation**

RIBONUCLEIC ACID (RNA)
A long chain nucleic acid involved in protein synthesis, consisting of repeating units containing adenine, cytosine, guanine and uracil. see **Deoxyribonucleic acid**

RICHARDSON NUMBER (Ri)
A measure of the relative rate of production of convective and mechanical energy. This number characterizes the relative importance of heat convection and mechanical turbulence.

RICKETTSIAE
Plant cells which resemble bacteria but are considerably smaller. They occur as intracellular parasites of ticks, lice, mites and fleas. Many have been shown to be parasitic.

RINGLEMANN NUMBER
Characterization of a visible black smoke in terms of opacity. Four charts consisting of grids of black lines are used: 0 is all white, 1 has 20% black area, 2 is 40%, 3 is 60% and 4 is 80%. 5 is all black. These charts were developed in the nineteenth century for use by smoke inspectors. The chart is held at eye level at a distance such that the chart lines merge into shades of gray and estimation of the opacity of the smoke plume in question is made. Although the Ringlemann chart is easy to use and requires no instrumentation, there are serious drawbacks. There is no strong quantitative relationship between Ringlemann number and emissions concentrations. Further, it is an esthetic and subjective measurement and has no direct bearing on many air pollution problems such as physiological effects and corrosion.

RIO DECLARATION
see **Conference on Environment and Development**

"RIP AND SKIP"
The term applied to many unscrupulous asbestos removal practices.

RISK ASSESSMENT

Characterization of potential adverse results resulting from exposure to hazardous substances. This may include estimates of uncertainties in analytical techniques.

RIVER

A stream of fresh water flowing in a natural watercourse.

RNA

see **Ribonucleic acid**

ROENTGEN (r)

The amount of gamma or X-radiation that will produce one electro-static unit of electricity in one cubic centimeter of dry air at standard temperature and pressure. This is equivalent to 1.61×10^{12} ion pairs per gram of air and corresponds to 83.8 ergs (8.4×10^{-6} J) of energy. see **Rad**. see **Rem**. see **Rep**

ROTATING BIOLOGICAL CONTACTOR (RBC)

Large (2–4 m diameter) plastic disks rotate slowly while half sub-merged in flowing wastewater. Biomass grows on the disk surface and substrate absorption occurs during submergence. Oxygen is absorbed when the biomass is in contact with the air. An RBC operates like a trickling filter.

ROTIFER

The simplest of all multicellular animals, its name comes from the rotating motion of the two sets of cilia on the head. Rotifers are strict aerobes and found only in waters with low organic content. These animals are considered indicator organisms for waters of low pollution load.

ROUGHNESS LENGTH

This is not a measure of the actual height of the boundary roughness elements. It is proportional to the size of eddies which can exist among the roughness elements.

ROUNDWORM

Also called ascaris, this disease is caused by a large intestinal roundworm. The ova can remain viable in the soil for a long time and control of the disease is by careful control of sewage and sewage sludge.

ROYAL COMMISSION STANDARDS (UK)

Sewage effluent standards for BOD (20 mg/L) and suspended solids (30 mg/L), sometimes called 30/20, promulgated by the Royal Commission on Sewage Disposal in 1915, have remained essentially unchanged.

RSA

see **Radioactive Substances Act**

RUBBISH

Wastes from homes, small businesses, etc., excluding garbage. Trash has this meaning in the US. In the UK rubbish refers to discarded material.

S

SAFE DRINKING WATER ACT (SDWA) (US)

This Act deals with the quality of tap water and specifies procedures for setting contaminant levels for public water systems. Also covered are underground injection of wastewater and federal development in sole source aquifers.

SALINE WATER

Water in which the total dissolved solids concentration is greater than 20 000 mg/L. Seawater contains about 3.5% dissolved solids (35 000 mg/L).

SALMONELLOSIS

Gastro-enteritis is the most common infection associated with the water cycle. The *Salmonella* bacterium is carried by humans and animals. Infection is by ingestion of large numbers of viable organisms in fecally contaminated food and water. Sewage sludge applied to grazing land can contain viable organisms for up to three weeks. The Commission of the European Communities has set restrictions of three weeks for grasslands after application of sewage sludge.

SALR

see **Saturated adiabatic lapse rate**

SALT WEDGE
When seawater intrudes into a tidal river, less dense outflowing water will overlie the denser oscillating saline water. Water exchange is usually from the salt to the fresh.

SAND FILTER
see **Mixed media filter**. see **Rapid sand filter**. see **Slow sand filter**

SANITARY LANDFILL
This is a carefully engineered method for disposal of solid waste. The waste is spread in thin layers, compacted and covered with a thin layer of earth at the end of the working period. Properly operated, this is an excellent means of solid waste disposal but suitable sites for sanitary landfills are becoming difficult to find in many areas. It is necessary to exercise care in operation of landfills. Without proper care, these become simply dumps.

SAPROPHAGES
Organisms which obtain their energy from dead or decaying materials.

SAR
see **Specific absorption rate**

SARA
see **Superfund Amendments and Reauthorization Act**

SATURATED ADIABATIC LAPSE RATE
A parcel of air saturated with water vapor will have a rate of energy release less than that of dry air. see **Adiabatic lapse rate**

SATURATION ZONE
That portion of an aquifer in which the pores are completely filled with water.

SBT
see **Segregated ballast tanks**

SCHISTOSOMIASIS
This waterborne disease is caused by a trematode worm that spends much of its life in a water snail but a part can be spent in a human or animal host. Humans become infected by swimming, wading or working

in freshwater. Reservoirs of the disease are warm blooded animals. Control involves eradication of the snail host and provision of adequate sanitary facilities.

SCHMUTZDECKE
The cover of material collecting on the surface of a slow sand filter. It was thought that this layer was necessary for straining of floc particles but it is now recognized that this putrescible mat is not necessary. The removal mechanism involves adsorption onto the surface of the filter medium and straining is not carried out to any significant degree.

SCHULZE-HARDY RULE
Precipitation of a colloid is achieved by addition of an electrolyte whose ion has a charge opposite to that of the colloid. Ability of ions to neutralize colloidal particle charge roughly 1 to 100 to 1000 for univalent, divalent and trivalent ions.

SCRUBBER
see **Gas absorber**

SDWA
see **Safe Drinking Water Act**

SEASAT
see **Earth Resources Technology Satellite**

SEAWATER
Major constituents of sea water are, in ppm (mg/L):

Na^+	10 600
K^+	1320
Mg^{+2}	406
Ca^{+2}	390
Cl^-	19 200
SO_4^{2-}	2700
HCO_3-	140
Br^-	65
F^-	1.3

pH of seawater is about 8.

SEAWEED

Marine macrophytes which require light, water and mineral ions to produce organic matter. Seaweeds are one of the major living sea resources.

SECCHI DISK

A simple means of measuring changes in transparency of water. A 20 cm diameter disk with alternating white and black quadrants is lowered into water until it is no longer possible to discern differences between the white and black quadrants. The depth is recorded and compared to test results from different times.

SECONDARY CONSUMERS

Animals which use herbivores for food.

SECONDARY PATHOGENS

Thermophillic fungi may infect individuals already infected by other pathogens or who are particularly susceptible to respiratory complaints.

SECONDARY POLLUTANTS

Formed as a result of some reaction in the atmosphere. The reaction may occur among any combinations of air pollutants and natural components of the ambient atmosphere.

SECONDARY SETTLING

In biological wastewater treatment the pollutant material is utilized as food by microorganisms. The food becomes part of the microbial cell and is removed in tanks which follow the biological process.

SECONDARY TREATMENT

Biological treatment of wastewater. see **Primary treatment**. see **Tertiary treatment**

SECOND-HAND SMOKE

see **Environmental tobacco smoke**

SECOND ORDER REACTION

One in which the rate of reaction is proportional to the square of the concentration of a reactant or the product of the concentrations of two reactants.

SEGREGATED BALLAST TANKS

These tanks, in which ballast water is carried in ships, have no physical connection to fuel or cargo tanks. As a result, water in these tanks is not contaminated, as is ballast water carried in fuel or cargo tanks.

SELECTIVE CATALYTIC REDUCTION (SCR)

A system for NO_x removal in which gaseous ammonia reacts with NO in the presence of oxygen to form nitrogen gas and water.

SELENIUM

A member of Group VIA of the periodic table, this metal occurs widely but unevenly in the earth's crust. Its concentration is about 4 micrograms per liter in seawater. It is toxic in elevated concentrations to humans and aquatic life. The maximum acceptable concentration is 10 micrograms per liter in drinking water.

SEPARATE SYSTEM

A sewer system which has separate pipes for conveyance of wastewater and storm water. see **Combined sewer**

SEPARATION

In management of solid waste it is necessary to separate various components of the waste stream. This can be accomplished by shredding and classifying by density in an air classifier. Hydropulping can accomplish the same result.

SEPTAGE

Contents removed from septic tanks, portable vault toilets, holding tanks, grease traps, very small treatment plants and semi-public facilities. These wastes are high in organics, grease and solids. Disposal is to municipal treatment plants.

SEPTIC PRIVY

see **Human waste**

SEPTIC TANK

A private wastewater disposal system used for some homes in rural areas. Wastewater is discharged to a tank, usually concrete, in which solids settle to the bottom where anaerobic decomposition takes place.

The overlying water flows to a subsurface leaching (tile) field. Suitability of the soil for percolation is the most important consideration.

SETTLEABLE SOLIDS
Solids suspended in water which will settle under quiescent conditions.

SEWAGE
The spent water supply of a community. Municipal sewage is about 99.95% water and 0.05% waste material. If the collection system is loose, there will be more water reaching the treatment plant than was supplied to the community due to infiltration of groundwater into loose pipe joints.

SEWER ORDINANCE
A regulation setting forth what may or may not be discharged to a collection system. Substances such as gasoline and hexavalent chromium are universally excluded. see **Prohibited discharges**

SHIGELLOSIS
Bacterial dysentery. *S. sommei, S. flexneri* and *S. dysentariae* are spread primarily by the path of anus to mouth. The most severe symptoms are associated with *S. dysenteriae.* Good sanitary practices are the best way to deal with Shigellosis.

SI
see **International System (of Units)**

SIC
see **Standard industrial classification**

SIERRA CLUB (US)
A non-profit organization founded at the end of the nineteenth century by influential interested citizens to deal with conservation issues.

SIEVERT (Sv)
see **Rem**

SIGMA-T VALUE
A shorthand way of expressing density of saline water. The density of sea water at 20°C and 35 parts per thousand of salt is 1.024785 gm/cm^3 This is written as σt = 24.785.

SILICOSIS

This occurs as a result of inhaling free silica below 7 microns in diameter. It is most common among workers in quarrying, mining, tunneling operations and sandblasting. Fibrotic lung lesions form and often calcify, producing restrictive lung disease.

SIP

see **State Implementation Plan sources**

SLOP TANK

A tank for holding contaminated water from cargo and fuel tank cleaning operations.

SLOW SAND FILTER

Water is allowed to pass slowly through a sand bed about 36 inches deep. When the filter becomes dirty and the filter runs become too long, the filter is taken out of service for cleaning. The upper one or two inches of the bed, on which almost all floc has been adsorbed, are discarded. Slow sand filters have been almost entirely replaced by rapid sand filters and mixed media filters. see **Mixed media filters**. see **Rapid sand filters**

SLUDGE

Solid material taken from wastewater by sedimentation or flotation. This can contain wasted microbial mass from biological treatment. It contains 95–98% water.

SLUDGE AGE

Ratio of the total active microbial mass in a treatment system to the active microbial mass withdrawn daily.

SLUDGE DIGESTION

Stabilization and disposal of organic solids removed from primary settling tanks and excess biological sludge from trickling filters and activated sludge units is accomplished by microbial decomposition under controlled conditions. The more common practice is anaerobic digestion but aerobic digestion has been carried out. Aerobic digestion produces a greater volume of end products than is produced in the anaerobic process. Raw sludge has a water content of 95% or more and is difficult to dewater. It has an unpleasant odor. Microorganisms in large tanks called sludge digesters reduce complex compounds to simpler substances and, in doing so, allow easier dewatering with associated volume reduction. Process temperature is about 95°F (35°C). A uniform temperature is achieved by use of external heat exchangers. Gases produced, principally methane, can be burned to supply process heat. Digested sludge can be dewatered on sand beds or by mechanical means. Disposal of digested sludge is a serious problem due to lack of acceptable disposal sites. Ocean dumping is no longer allowed. Sludge can be incinerated but this introduces the possibility of air pollution.

SLUDGE VOLUME INDEX (SVI)
The ratio of the volume occupied by 1 gram of mixed liquor suspended solids after 30 minutes settling compared to the volume occupied by 1 gram on a dry weight basis. The sludge volume index of a well settling activated sludge is between 50 and 100 and bulked sludge, an undesirable situation, has an SVI of 200 or greater.

SMOG
An artificial fog produced by photochemical reactions in a polluted atmosphere close to the ground.

SMOG CHAMBER
A relatively large photochemical reaction vessel in which chemistry occurring in the environment can be simulated.

SMOKE
A dispersion of solid particulate matter in air. Aerosols and particles in smoke carry a charge and neutralization of charge is necessary for their removal. This can be accomplished by passage through an electrostatic precipitator.

SMOKE SHADE
see **Ringlemann number**

SOAPS
Materials for cleaning which are derived from fats and oils by treatment (saponification) with hydroxide. In hard waters the hardness causing ion (calcium or magnesium) may precipitate the soap.

SODIUM FLAME TEST

A test for high efficiency air filters in an NaCl atmosphere. The flame burns more brightly on the dirty side of the filter than on the clean side and filter efficiency is determined by comparing brightness of flames on the two sides.

SOIL

Weathered rock debris and humus which can support plant and animal life. Organic and mineral matter accumulate in the A-horizon. On the B-horizon the minerals and clay washed down from the overlying A-horizon are found. Another name for B-horizon is subsoil. Below this is the C-horizon in which there are not many soil flora and fauna.

SOIL HORIZON

see **Soil**

SOIL (COMPOSTING) PRIVY

see **Human waste**

SOIL VAPOR EXTRACTION (SVE)

One of the problems in bioremediation is adequate air for metabolic processes of aerobic organisms. Drawing a vacuum through unsaturated soils allows a flow of air through subsurface soils. Soil permeability is significant in efficiency of air supply. see **Bioremediation**

SOLAR CELL

see **Photovoltaic cell**

SOLAR ENERGY

Conversion of the energy in the sun's rays into energy for direct use. This is regarded as a promising energy source for Third World countries. In general, in these countries, it is felt that generation should be on-site because of the lack of transmission systems. Home heating by solar power is becoming more common in developed countries. Engineering problems, including the requirement of large areas for solar panels, can be solved. The main difficulty with solar energy utilization is dedication at the government level.

SOLAR RADIATION

Energy reaching the earth from the sun. At the outer surface of the atmosphere, intensity is a closely constant value of 1400 watts per square meter (2 calories per square centimeter per minute). 98% of the energy is in the wavelength range 0.2–4.5 nm. Of this, 40–45% is in the visible wavelength range 0.4–0.7 nm. About 1/3 is reflected back into space and only about 50% of the radiation reaches the earth. The figure is greater for dry regions and less for tropical regions. see **Albedo**

SOLE-SOURCE AQUIFER (US)
Under the Safe Drinking Water Act, this is an aquifer that is the only source or potentially the only source of drinking water for an area.

SOLIDS
Material remaining as residue after cooling when water is evaporated at 103°C.

SOLIDS RETENTION TIME
The average residence time of solids (sludge) in a treatment process. This an is important parameter in anaerobic sludge treatment.

SOLID WASTE
For regulatory purposes (US), solid waste is any discarded material, including liquids and gases in containers.

SOLUBILITY
The extent to which one chemical species (solute) dissolves in another, usually liquid, species (solvent). Gases become less soluble with temperature increase but liquids and solids become more soluble.

SOLUTE
see **Solubility**

SOLUTION STRENGTH
A molar solution is one in which one gram-molecular weight is dissolved in one liter of total solution. An equivalent solution has one gram of replaceable hydrogen per liter or can react with one gram of hydrogen (eq/L = valence × mol/L). see **Standard solution**

SOLVENT
see **Solubility**

SORBENT
A buoyant substance which can preferentially absorb oil, this can make cleanups easier by concentrating the oil. Materials range from natural, such as straw and sawdust, to synthetics such as polyurethane and polyester shavings.

SOUND
Audible pressure waves in air with frequencies from 20 Hz to 20 000 Hz.

SOUND LEVELS
Sound energy is proportional to mean-square sound pressure. The following sound levels are A-weighted:

Sound level (A-weighted)	Source or criterion
140	Threshold of pain
122	Supersonic aircraft
112	707 airliner
100	Air hammer
90	OSHA 8 hour limit (US)
80	Loud speech
65	Daytime limit, typical ordinance
60	Normal conversation
50	Night time, typical ordinance
35	Acceptable for sleep

SOUND PRESSURE
The difference between instantaneous absolute pressure and ambient pressure.

SOUTHERN OSCILLATION
The seasonal shift in atmospheric surface pressure between the Australian Indian Ocean and southeastern Pacific. Variations in this shift caused by warming of shore waters off the Peruvian coast is thought to give rise to El Nino phenomena such as abnormal rainfall events on a global scale.

SO$_x$
Oxides of sulfur. SO_2 and SO_3.

SPECIFIC ABSORPTION RATE

The rate at which electromagnetic radiation is absorbed in a medium per unit mass of medium, expressed as watts/kg.

SPECIFIC GROWTH RATE

In microbial culture systems, the rate of cell growth is proportional to the concentration of microorganisms. The proportionality constant is the specific growth rate.

SPECIFIC HEAT

The amount of heat required to raise the temperature of a unit quantity of a substance one degree.

SPECIFIC UTILIZATION

see **M/F ratio**

SPECTROPHOTOMETER

An instrument for measuring attenuation of incident light passing through a solution. It is widely used in analytical work and is suitable for measuring low concentrations. All spectrophotometers have an energy source, an energy spreader for choice of wavelength and an energy detector.

SPENT NUCLEAR FUEL

Irradiated nuclear reactor fuel after removal from service and before reprocessing.

SPHERICITY

A measure of how closely a grain approaches the shape of a perfect sphere. This is important in determining flow rates through porous media. see **Kozenty equation**

SPIROCHETES

These are not true bacteria but are unicellular and reproduce by binary fission. The most important is *Treponemia palladium*, the causative organism of syphilis.

SPORE

Microorganisms surround themselves with a tough polysaccharide layer when exposed to adverse environmental conditions and this protects the cell nucleus. The polysaccharide layer is dissolved when conditions improve.

SPOROCIDE
A chemical agent which destroys bacterial spores.

SRT
see **Solids retention time**

SS
see **Suspended solids**

STABILIZATION
When applied to hazardous wastes, this involves fixing the material with fly ash and lime to produce a solid unleachable end product suitable for land disposal.

STANDARD DEVIATION
A measure of the dispersion of the data about the mean value. The square of the standard deviation is the variance.

STANDARD INDUSTRIAL CLASSIFICATION (US)
Twenty broad industry types are assigned two-digit identification numbers. More specific classification is achieved through further assignment of one or two digits. These numbers indicate further subdivision within the broad category: 20 is food, 202 is dairy products and 2022 is cheese.

STANDARD METABOLIC RATE
The level of oxygen consumption which is just necessary to maintain vital functions of a resting fish.

STANDARD METHODS (FOR THE EXAMINATION OF WATER AND WASTEWATER)
A collection of generally accepted analytical methods for examination of water and wastewater. Different analytical methods can give somewhat different results when applied to the same solution. Periodically, committees of experts drawn from professional groups agree on methods for analysis. Professional societies sponsor the final publication.

STANDARD SOLUTION
One whose strength or reacting value is known. A normal solution is one which contains one equivalent per liter (1 eq/L = 1N).

STANDING CROP

The measured part of biological production per unit area or unit volume physically present and not lost through respiration.

STATE IMPLEMENTATION PLAN SOURCES (US)

Pollutant sources built prior to establishment of new source standards. Emissions levels are determined on an individual basis and depend on the area air quality.

STEP AERATION

A variation of the activated sludge process. Wastewater is introduced at points along the aeration tank in order to even out the oxygen demand and keep the sludge reaerated in the presence of substrate (food). As a result, the process is more active biologically and does not go into the endogenous phase near the end of the aeration tank.

STERILIZATION

The process of destroying all microbial life.

STERLING ENGINE

Combustion of fuel is external to the engine and heated fluid (air) is led to the piston for the power stroke. A steam engine is also an external combustion engine.

STILBESTEROL

see **Diethylstilbesterol**

STOCHASTIC MODEL

One which includes probability in formulation.

STOCHASTIC PROCESS

One which includes a random element in its description.

STOKES DIAMETER

Diameter of a particle having the same gravitational settling velocity as the particle under consideration.

STOKES LAW

The frictional force acting on a spherical particle moving at terminal velocity is a function of the velocity, sphere diameter and fluid viscosity.

STORAGE COEFFICIENT
The amount of water released from a column of an artesian aquifer through one square foot when the head acting on the aquifer falls one foot.

STRATOSPHERE
The upper part of the atmosphere, extending from a height of 8–16 km to 75–80 km above the surface.

STREETER-PHELPS EQUATION
see **Oxygen sag equation**

STRIP MINING
Removal of overlying soil (overburden) in order to mine minerals near the surface. Restoration of the area after cessation of activities is a serious issue.

STRONTIUM
A hard white metal of Group IIA of the periodic table occurring between barium and calcium. It is a "bone seeker" when ingested and radioactive strontium 90 can remain in the body for an extended period.

S. (SALMONELLA) TYPHOSA
see **E. typhosa**

SUBLITTORAL ZONE
The sea-shore zone lying immediately below the littoral (inter-tidal) zone.

SUBSTRATE
In waste treatment, the food utilized by microorganisms in reducing the waste to a more stable end product. In microbiology, the molecule which reacts with an enzyme or is used for cell growth. see **Green–Stumpf theory**

SUBSTRATE REMOVAL VELOCITY
see **M/F ratio**

SULFATE (SO_4^{-2})
A major anion occurring in natural waters, it has a cathartic effect on humans and forms hard scales in boilers and heat exchangers. When

reduced, sulfate gives hydrogen sulfide, responsible for odors and corrosion of sewers.

SULFUR BACTERIA
These obtain energy for life processes through oxidation of reduced forms of sulfur such as hydrogen sulfide. Sulfur bacteria, *Thiobacillus,* can exist at pH values below 1.0.

SULFUR DIOXIDE
A colorless gas whose odor and taste can be detected at low concentrations and which has a pungent, irritating odor above 3 ppm. Primary sources of SO_2 and SO_3 are from burning coal and crude oil for space heating and power generation. SO_2 is a lung irritant but the sulfuric acid aerosol formed when SO_2 is oxidized to SO_3 is most damaging to health. The aerosols are usually less than 2 microns in diameter and penetrate the lung passages. SO_2 can damage crops, plants and trees.

SUMMATION
Agents with similar pharmacological actions produce a response which is the sum of the individual agents.

SUPERFUND
see **Comprehensive Environmental Response, Compensation and Liability Act**

SUPERFUND AMENDMENTS AND REAUTHORIZATION ACT (SARA) (US)
Signed into law in 1986, SARA increased the scope of CERCLA. The trust fund was increased, assessment of potential sites for inclusion in Superfund's National Priority List was accelerated and mandatory deadlines set for remedial investigations and actions. Statutory authority was given for settlement agreements and criminal penalties were increased.

SUPEROXIDATION
Application of a strong oxidizing agent such as chlorine or ozone to wastewater sludge in order to stabilize organic material and facilitate dewatering for volume reduction.

SURFACE MINING CONTROL AND RECLAMATION ACT (US)

This provides for federal regulation of private surface coal mining on private lands and strip mining on public lands. Administered by the Office of Surface Mining in the Department of the Interior, this Act prohibits surface mining in environmentally sensitive areas.

SURFACTANT

A compound that can reduce the surface tension of a liquid and that contains both water compatible (hydrophilic) and oil compatible (lipophilic) groups.

SUSPENDED SOLIDS

Matter which is retained on a suitably prepared Gooch crucible. Results are expressed in mg/L and give an indication of the sludge beds to be expected at wastewater outfalls. Upon heating at an elevated temperature of 600°C, suspended solids can be differentiated as to fixed and volatile solids. The residue from the test is called fixed solids and the material driven off is a measure of the organic (volatile) component.

SUSTAINABLE YIELD

Material which can be harvested from an ecosystem over an extended period without harming the ecosystem.

SUTRO WEIR

Also called a keyhole weir because of its shape, this is a device for measuring volumetric flow in an open channel.

SUTTON EQUATION

The basic expression on which almost all atmospheric diffusion models are based, this is Gaussian in form. Downwind pollutant concentration is related to continuous source strength, wind speed and atmospheric stability parameters.

SVE

see **Soil vapor extraction**

SVI

see **Sludge volume index**

SYMBIOSIS

Two organisms cooperate for mutual advantage.

SYNDETS
see **Synthetic detergents**

SYNECOLOGY
The study of entire communities in relation to their environment. see **Autoecology**. see **Ecology**

SYNERGISM
The case where the effects of two or more substances together are greater than the sum of the individual effects.

SYNGAS
see **Fischer–Tropsch Process**

SYNTHETIC DETERGENTS
Commonly called syndets, these were introduced as substitutes for soap. The major advantage is that they do not form insoluble precipitates with soap. Two serious problems with environmental significance were foaming in aeration tanks and in receiving waters at points of turbulence and eutrophication. Replacement of ABS (alkyl benzene sulfonate) by more easily degraded LAS (linear alkyl sulfonate) solved the foaming problem. A key trigger to explosive algal blooms was a critical nitrate nitrogen to phosphate phosphorous ratio. Detergent packages contained about 70–80% phosphate as builders for water softening. Phosphates have now been replaced in commercial applications as a result of public outcry.

SYNTHETIC ORGANIC COMPOUNDS
Included are insecticides, fungicides and pesticides. These are not found naturally and are designed to be toxic to a specific target. The important groups are organochlorine, organophosphorous and organotin. see **Organochlorine insecticides**. see **Organophosorous**. see **Organotin**

T

TAEE
see **Reformulated gasoline**

TAME
see **Reformulated gasoline**

TAPEWORM
The beef tapeworm, *Taneia saginata,* has cattle as the intermediate host and humans as the final host. Cattle are infected by grazing on pasture recently spread with infected sewage or drinking infected water. Human infection is from ingesting the larval stage through raw or poorly cooked beef. Control of tapeworm is primarily through high standards of meat inspection.

TASTES AND ODORS
The olfactory and taste senses are closely connected and it is difficult to differentiate the two responses when exposed to a strong stimulus. Many sensations ascribed to taste are actually odors.

TCE
see **Trichloroethylene**

TDH
see **Total dynamic head**

TEMPERATURE
A measure of the kinetic energy of the molecules of the material.

TEMPERATURE COEFFICIENT
Activity of a disinfectant varies with temperature. Change in activity is quantified as a ratio of disinfection rate constants at two temperatures and is found to be fairly constant over a wide range of temperatures. The most commonly used comparison is for a 10°C change.

TEMPERATURE INVERSION
see **Atmospheric inversion**

TERATOLOGY
The study of congenital malformation which can be recognized at birth or soon thereafter. Viral infections, ionizing radiation and chemicals can be teratogenitic. Thalidomine was one of the most sensational teratogens.

TERTIARY TREATMENT
Any wastewater treatment, in addition to conventional secondary treatment, by which further removals of impurities are achieved. This does not include chlorination of secondary effluent. Some tertiary treatment processes are sand filtration, microstrainers, oxidation ponds, foam separation, activated carbon absorption, chemical clarification and ion exchange. see **Primary treatment**. see **Secondary treatment**. see **Water reuse**

THEIS EQUATION
A real world expression describing drawdown in wells which have not reached equilibrium. It is the most widely used analytical method for radial flow into a well. see **Well function**

THERMAL POLLUTION
Two problems are created when cooling waters are discharged to a watercourse. Temperature of the receiving water can be raised significantly and this will decrease the amount of dissolved oxygen which the water can hold while, at the same time, biological activity is stimulated, further lowering the dissolved oxygen.

THERMAL TOLERANCE ZONE
The temperature range in which 50% of a sample of fish of similar size and health survive.

THERMOCLINE
The point of maximum vertical temperature change in a lake or reservoir.

THIN EGGSHELL SYNDROME
Eggs laid by some birds which had ingested significant amounts of chlorinated pesticides were found to lack structural stability during incubation. The eggs cracked and survival of certain species was threatened.

THIOALCOHOL
see **Mercaptans**

THRESHOLD LIMIT VALUE (TLV)
The average 8 hour occupational exposure limit. Instantaneous values may be greater than this value but the 8 hour average must be the TLV or less. These values are thought to be safe exposure for the working lifetime.

TIDAL POWER
Tides occur as a result of interaction between the earth and the moon. Construction of dams permits harnessing of energy due to periodic water level differences for production of electric power. This means of energy production, while theoretically attractive, has many associated problems and has not been used widely.

TIDE
Periodic rise and fall of the ocean surface. The forcing is derived from the gravitational attraction of the moon, sun and other planets. The moon plays the dominant role due to its close proximity.

TILE FIELD
Effluent from a septic tank is distributed for disposal into the soil by a system of perforated field tiles.

TITLE 40 (US)
This is the part of the US Code of Federal Regulations (CFR) dealing with protection of the environment. It is updated annually by the Federal Register.

TLV
see **Threshold limit value**

TON

In air conditioning, 12 000 Btu/hr (3517 J/sec).

TOTAL DEPOSITION

Transfer of gas particles and precipitation from the atmosphere to the ground. The sum of wet and dry deposition.

TOTAL DYNAMIC HEAD (TDH)

The sum of the energy loss due to pipe friction and change in elevation (static lift).

TOXICANT

see **Toxic substance**

TOXICITY

Reaction of a substance or combination of substances to deter or inhibit metabolic processes without completely altering or destroying a species. Toxicity is a function of the nature of the substances, concentration, sensitivity of the target species, exposure time and environmental conditions. In some cases, a species can, over time, develop tolerance to an inhibitory concentration and become acclimated.

TOXICOLOGY

The science of poisons. The scientific discipline dealing with quantification of injurious effects on living systems which result from chemical and physical agents that cause alteration in a cell or organ function or structure.

TOXIC SUBSTANCE

For there to be a toxic situation, three elements must be present: (1) a chemical or physical agent capable of producing a response; (2) a biological system with which the agent may react to produce a response; (3) a response which can be considered deleterious to the biological system.

TOXIC SUBSTANCES CONTROL ACT (TSCA) (US)

Under terms of this legislation, testing of new and existing chemicals that are potentially toxic is required. Conditions are placed on manufacture, distribution and usage of a chemical if it poses unreasonable risk to human health or the environment.

TOXIN
A poisonous substance of animal or vegetable origin.

TRACHEOBRONCHIAL
The portion of the respiratory tract from the bronchi down to terminal bronchioli.

TRADE WASTE
see **Industrial waste**

TRANSFER STATION
A facility that receives solid waste from smaller collection vehicles for transfer to larger vehicles, which then convey the waste to the ultimate treatment, disposal or recycling site.

TRANSPIRATION
Loss of water from plant surfaces to the atmosphere.

TRANSPORT VELOCITY
Velocity in a duct sufficient to keep collected particles airborne.

TRANSURANIC WASTE
Waste contaminated with alpha emitting radionuclides with atomic numbers greater than 92, half lives greater than 30 years and which have activities greater than 100 nanocuries per gram (3700 Bq/g).

TRASH
see **Rubbish**

TREATABILITY STUDIES
Some wastes, particularly those of industrial process origin, require special study to develop suitable treatment. Usually such treatability studies are conducted at the bench scale level.

TREMATODES
see **Helminths**

TREND
The slope of the line relating variables.

TRICHLOROETHYLENE (TCE)
A common vapor degreasing solvent often found as a groundwater contaminant.

TRICKLING FILTER

Described most simply, this aerobic wastewater treatment process allows previously settled wastewater to trickle slowly over stones or other media on which a biological slime has developed. Waste material in the water is used as food by the organisms in the filter slime. Usually, the wastewater is discharged from rotating arms above the filter. Treated water is collected in underdrains and lead to a secondary (humus) tank in which sloughed off filter slime is collected. A portion of the filter effluent is recirculated in order to smooth out the variations in waste (food) supply to the process. Variations of the basic process include low, intermediate, high and super high rate filters and rotating biological contactors. Most modifications have resulted from specific operating problems. see **Rotating biological contactors**

TRIHALOMETHANES

Bromoform, chloroform, chlorodibromomethane and dichlorobibromomethane. These result from interaction between chlorine-based disinfectants and naturally occurring organic compounds such as humic and fulvic acids.

TRITIUM

An unstable radioactive isotope of hydrogen with a mass of three and a half-life of 12.6 years used in fusion power research and as a biological tracer.

TROPHIC LEVEL

The position of an organism in a community based on its feeding requirements.

TROPICAL RAIN BELT

Clouds near the equator move north during part of the year, bringing monsoons to normally arid regions. If these clouds do not move north, the arid regions suffer droughts and food shortages soon develop.

TROPICAL RAIN FOREST

see **Rain forest**

TROPICAL SOILS

Predominant soils of humid tropical regions are oxisols, ultisols and alfiols. These are highly weathered, have low cation exchange capacity, low plant nutrient reserves, low water holding capacity and are prone to compaction and rapid erosion. These can have toxic levels of manganese and aluminum.

TROPOSPHERE
The lower level of the atmosphere, extending to about 8 km at the poles and 16 km at the equator.

T(TAENIA) SAGINATA
see **Tapeworm**

TSCA
see **Toxic Substances Control Act**

T,T,T,O
Time, temperature, turbulence and oxygen. These are the elements necessary for a good combustion process. There must be sufficient time for burning, a sufficiently high temperature must be attained, there must be adequate turbulence for good mixing and sufficient oxygen for complete combustion. Too much air reduces the combustion temperature.

TUNDRA
The treeless region above 60°N latitude and south of the polar ice. Drought, low temperatures, cold dry winds and permafrost are the factors which limit tree growth.

TURBIDITY
Interference with passage of light through water or air. In air, this can be due to haze, smoke or other particles. In water, it can be caused by suspended particles ranging in size from colloidal to coarse suspensions. Turbidity is important because of esthetics, filterability, efficiency of disinfection and in water bodies can limit the depth at which photosynthetic activity can occur.

TURBULENT FLOW
This regime occurs at Reynolds numbers above 4000. For a pipe the characteristic length is the diameter. Turbulent flow is characterized by a flat velocity profile.

TYNDALL EFFECT

Colloidal particles have dimensions greater than some wavelengths of white light and interfere with light passage. A beam of light passing through a colloidal suspension is reflected and is visible to an observer at right angles to the light path. This phenomenon is also important in observing behavior of aerosols.

TYPHOID FEVER

An enteric disease spread by waste contaminated water and food. Prevention of the spread of typhoid fever was formerly the underlying factor in water sanitation practice but now protection of the oxygen resources of the receiving water achieves the same objective.

U

ULTISOLS
see **Tropical soils**

ULTRASOUND
Sound with a frequency above the audible range.

ULTRAVIOLET RADIATION
The sun is the major source of ultraviolet radiation. Absorption by the ozone layer allows only wavelengths greater than 290 nm to reach the earth's surface. Biological effects of UV include erythema (reddening of the skin) and skin cancer. The UV spectrum has been subdivided into three segments. UVA (320–400 nm), UVB (280–320 nm) and UVC (200–280 nm). UV is increasingly active with decreasing wave length. UVA is the least active while UVC is used for its bactericidal properties.

UNCLOS
see **Law of the Sea**

UN CONVENTION ON THE LAW OF THE SEA, 1982
see **Article 60**

UNIFORMITY COEFFICIENT
Sieve analysis of sand gives a percentage distribution of mean diameters. A measure of the uniformity of the sand is the ratio of the

diameter which is greater than 60% (d_{60}) of the sample to the diameter which is greater than 10% (d_{10}).

UNIT HYDROGRAPH
A temporal plot of volumetric runoff in a stream, the area of which (less base flow) represents one inch of runoff from the catchment area. The unit hydrograph is used to predict flows resulting from various rainfall combinations applied to the area.

UNITIZED CARGO CARRIERS
Ships which are designed for transport of standard containers, wheeled vehicles and general break bulk cargo. These ships have adequate space for ballast tanks and ballasting of fuel and cargo tanks is not necessary.

UNIT RISK VALUE
The increased lifetime cancer risk occurring in a population in which all individuals are exposed continuously from birth until age seventy to a concentration of 1 microgram per cubic meter of a pollutant in the air they breathe.

UPWELLING
Wind acting along a coast, with the shore to the left in the northern hemisphere and to the right in the southern hemisphere, will cause transport of surface water in the offshore direction. The water will be replaced by nutrient rich bottom water from offshore. The colder bottom water transported to the shallow inshore region can have significant effect on climate. see **La Nina**

URANIUM
A dense lustrous metal resembling iron, the atomic number is 92 and the atomic weight is 238.03. Used in nuclear power production and weapons, there are 15 known isotopes. U-234, U-235 and U-238 exist in Nature. While uranium poses the greatest threat due to radiation, it is also toxic to about the same degree as arsenic.

URANIUM TAILINGS RADIATION CONTROL ACT (US)
This 1978 law directed the Department of Energy to provide for stabilization and control of uranium mill tailings from inactive sites.

UV
see **Ultraviolet radiation**

V

VADOSE ZONE
see **Phreatic water**

VAN'T HOFF RULE
A reaction rate will double for a 10°C rise and will fall to one half the value for a 10°C drop. This rule applies in the normal range of biochemical reactions and is sometimes written $Q_{10.}$ see $\mathbf{Q_{10}}$

VAPOR PRESSURE
Pressure exerted by molecules in the vapor state in a closed container in equilibrium with molecules of the substance in the liquid state.

VARIANCE
see **Standard deviation**

VECTOR
An organism which carries disease from one host to another. Common vectors are anopheles mosquitoes, common house flies, rats, lice, fleas and ticks.

VENTILATION FACTOR
The product of the thickness of the mixed layer and the average wind speed in the mixed layer.

VIBRIO COMMA (CHOLERA)
The causative organism of cholera.

VIENNA CONVENTION (FOR PROTECTION OF THE OZONE LAYER)
This outlined responsibilities of states for protection of human health and the environment against adverse effects of ozone depletion. It established the framework on which the Montreal Protocol was negotiated. see **Montreal Protocol**

VIRION
see **Virus**

VIRUS
The smallest plant cells known, viruses are intracellular parasites which derive their nutrients from the host organism. They are considered by some to be pure organic compounds which have the power to reproduce within the host. Viruses which are parasitic to bacteria are known as bacteriophages. Viruses must be observed by means of the electron microscope.

VISIBLE EMISSIONS (VE)
Measurement of plume opacity. A method for measurement is given under 40 CFR Part 60 (US).

VISCOSITY
Resistance to fluid shear.

VITRIFICATION
Immobilization of radioactive waste in a glasslike solid.

VLLW
see **Radioactive waste**

VOC
see **Volatile organic compounds**

VOLATILE ACIDS
These are low molecular weight short chain fatty acids (acetic, propionic, butyric, etc.) important in anaerobic sludge digestion. It has

been found that malfunctioning digestion processes will have elevated concentrations of volatile acids on the order of 6000 mg/L.

VOLATILE ORGANIC COMPOUND (VOCS)

These are man-made substances, some carcinogenic, which may contaminate groundwater supplies. They are usually found in industrial settings where such materials as solvents, degreasers and cleaning fluids have been discarded improperly. Conventional water treatment operations do not have significant effect on VOCs and it is necessary to treat these by packed bed aeration (air stripping) or activated carbon.

VOLATILE SUSPENDED SOLIDS

Also called volatile solids, these are suspended solids driven off when the sample is heated to 600°C. Volatile suspended solids are a measure of the organic load to be handled in a biological treatment system.

VOLUMETRIC LOADING RATE

see **Hydraulic loading rate**

VSS

see **Volatile suspended solids**

WARBURG RESPIROMETER
An instrument for measuring oxygen uptake by microorganisms.

WASHOUT
Chemicals incorporated into falling precipitation by below-cloud scavenging.

WASTEWATER COAGULATION
The most common coagulants used in chemical treatment of wastewater are aluminum sulfate, ferrous sulfate (copperas), lime and ferric chloride. All supply divalent or trivalent ions for neutralization of charge. A new approach is electrocoagulation. Coagulating ions are supplied by electrolytic oxidation of sacrificial electrodes.

WATER BALANCE
A method of assessing water resources in an aquifer, catchment area or larger areas. Evaluation of water supplies (recharge) is made against uses (abstraction).

WATER-BASED DISEASES
An essential part of the life cycles of infecting agents of these maladies takes place in an aquatic animal and infection can take place with or without ingestion of water. Included are bilharziosis, dracunculosis, onchalersosis, philariosis, schistosomiasis and threadworm.

WATER-BORNE DISEASES
The most common diseases transmitted by ingestion of contaminated water and food are amoebic dysentery, bacillary dysentery, cholera, gastroenteritis, giardiasis, hepatitis, leptospirosis, paratyphoid fever, salmonellosis and typhoid fever.

WATER BUDGET
see **Water balance**

WATERCRAFT WASTE DISPOSAL
see **Holding tank**. see **Incinerating toilets**. see **Maceration–disinfection device**. see **Recirculating toilets**

WATER CYCLE
see **Hydrologic cycle**

WATER RELATED VECTORS
Some diseases are spread by bites of insects which live and breed close to water. Among these are dengue fever, encephalitis, filariasis, hemorragic fever and malaria.

WATER REUSE
Further utilization of water while still under control of the first user. The degree of treatment necessary will depend on the further use to which the water will be put.

WATERSHED
A geographic area in which water, sediment or other materials will drain to a common outlet such as a river, stream, lake or swamp.

WATER TABLE
The upper limit of the saturation zone of an aquifer.

WATT
One joule per second.

WEATHERING
The mechanical (physical) fracturing or chemical decomposition of rocks by natural agents, temperature change, freezing and thawing and

action of plant roots. Chemical weathering results from reaction of the rock with fresh water.

WEBER-FECHNER LAW

A stimulus must increase geometrically (exponentially) if the response is to increase arithmetically (linearly). In the literature this is sometimes called Weber's Law.

WEIL'S DISEASE

see **Leptospirosis**

WELL FUNCTION

A definite integral occurring in solution of the non-equilibrium case of well drawdown. The functions have been tabulated and are also used in graphical solutions. see **Theis equation**

WET DEPOSITION

(1) The amount of material removed from the atmosphere by rain, snow or other precipitation. (2) The process of transferring gases, liquids and solids from the atmosphere to the ground during a precipitation event.

WET-DRY SAMPLING

Use of two sampling containers, one of which is exposed automatically during dry periods while the second is open only during precipitation.

WETLAND

Low lying land which has the water table at or near the surface. Wetlands may be fresh, brackish or saltwater. Wetlands are important habitats and nesting areas for many species. Many wetlands are being lost through development.

WET-ONLY SAMPLING

Use of an automated sampler which is open only during precipitation events.

WET SCRUBBER

A finely atomized stream of liquid used to capture particulates and gaseous pollutants from a gas stream.

WHO
see **World Health Organization**

WILD AND SCENIC RIVERS ACT (US)
Enacted in 1968 and since amended several times, this statute established a system of wild and scenic rivers. Eligibility criteria were set up for wild, scenic and recreational rivers. Joint administration is by the Secretaries of the Interior and Agriculture.

WILDERNESS ACT (US)
Lands under this Act must be administered to preserve the wilderness character and be devoted to scenic, recreational, historic and similar uses.

WILDLIFE SPECIES PROTECTION (US)
Federal legislation dealing with protection of particular species and habitat acquisition programs include the Marine Mammal Protection Act, the Wild and Free-Roaming Horses and Burros Act, the Bald Eagle Protection Act, the Migratory Bird Hunting Stamp Act and the Water Bank Act.

WINDROWS
Stacks into which composting materials are placed.

WINKLER METHOD
This is the standard analytical method for determination of dissolved oxygen. Several modifications of the basic method have been introduced to account for interfering substances.

WINOGRADSKY COLUMN
A laboratory model for studying microbial population development in natural environments. An important application is testing for biodegradability capacity of a water body or soil.

WORLD BANK
see **International Bank for Reconstruction and Development**

WORLD CONSERVATION STRATEGY (WCS)
Sponsored by the United Nations Education, Scientific and Cultural Organization (UNESCO), the Food and Agriculture Organization

(FAO), the International Union for Conservation of Nature and Natural Resources (IUCN) and the World Wildlife Fund (WWF), this 1980 policy document presented an integrated strategy for global conservation.

WORLD HEALTH ORGANIZATION (WHO)

The Constitution of the WHO defines health as "A state of complete physical, mental and social well-being and not merely the absence of disease or infirmity." The World Health Organization is a part of the United Nations and is charged with dealing with all aspects of health. Based in Geneva, it undertakes cooperation on health matters on a global scale and is able to transcend national boundaries.

WORLD STANDARD POPULATION

A statistical tool employed by epidemiologists for comparison of disease and mortality rates.

XENOBIOTIC COMPOUNDS

Those compounds which do not occur naturally.

X-RAYS

Also called roentgen rays, these are electromagnetic waves similar to gamma rays but having somewhat longer wavelengths. Shielding requires dense materials. In medical applications the eyes and gonads are the most susceptible structures.

Y

YU-CHENG INCIDENT
An outbreak similar to the Yusho disease of 1968 occurred in Taiwan in 1978. In both cases the rice oil used for cooking was contaminated by PCBs leaking into the food product from a faulty heat exchanger.

YUSHO DISEASE
Also called rice oil disease. A group of patients in Japan had ingested rice oil contaminated with Kanechlor, a chlorinated biphenyl. The symptoms were similar to chloracne. About 1800 persons were affected over a ten year period. The rice oil, used in cooking, had been contaminated by a PCB which had leaked from a heat exchanger. A similar incident occurred in Taiwan ten years later and is known as the Yu-Cheng incident.

Z

ZENO'S PARADOX

The problem of the time required to reach a wall if, in a given time interval, half of the remaining distance to the wall is traversed. This of interest because it is a description of a first order reaction.

ZEOLITES

These are also called molecular sieves because of their intricate pore structure. Both naturally occurring and synthetic zeolites function as ion exchangers. Early zeolites were natural, the most widely used being glauconite. see **Ion exchange.**

ZETA POTENTIAL

A measure of the strength of the charge on a colloid and the distance to which the effect extends. Strong zeta potential is associated with stable colloidal suspensions. Coagulation is aided by lowering zeta potential. This is accomplished by reducing the strength of the net charge on the particle.

ZINC

A widely occurring bluish-white metal resembling magnesium chemically. Zinc is an essential element in human and animal health but limits are placed on concentrations of zinc discharged to receiving waters in order to protect aquatic life. Zinc fumes can cause influenza-like symptoms in humans.

Z.(ZOOGLEA) RAMIGERA

This is thought to be the most important organism in the activated sludge process.

Acronyms and Abbreviations

AA	Acrylic Acid
AA	Atomic Adsorption
AAAS	American Association for the Advancement of Science
AABW	Antarctic Bottom Water
AAC	Antarctic Convergence
AAES	American Association of Engineering Societies
AAPS	American Association of Pharmaceutical Scientists
AASHTO	American Association of State Highway and Transportation Officials
AAT	Aspartate Amino Transferase
AAU	Assigned Amount Unit
ABET	Accreditation Board for Engineering and Technology (US)
ABGC	Air Blown Gasification Cycle
ABIH	American Board of Industrial Hygienists
ABL	Atmospheric Boundary Layer
ABS	Acrylonitrile-Butadiene-Styrene
ABS	Alkyl Benzene Sulfonate
ABW	Arctic Bottom Water
AC	Activated Carbon
ACBE	Advisory Committee on Business and the Environment
ACC	American Chemistry Council
ACC	Antarctic Circumpolar Current
ACE	Association for the Conservation of Energy (UK)

ACEEE	American Council for an Energy-Efficient Economy
AcFt	Acre-Feet
ACGCM	Atmospheric Coupled General Circulation Model
ACGIH	American Conference of Governmental Industrial Hygienists
ACH	Acetone Cyanohydrin
ACH	Air Changes Per Hour
ACM	Asbestos Containing Materials
ACNFP	Advisory Committee on Novel Foods and Processes (UK)
ACOPS	Advisory Committee on (Prevention of) Pollution of the Sea
ACS	American Chemical Society
ACSH	American Council on Science and Health
ACSNI	Advisory Committee on the Safety of Nuclear Installations (UK)
ACT	Alternative Control Technology
ADB	Asian Development Bank
ADCP	Acoustic Doppler Current Profiler
ADEME	Agency for Environment and Energy Management (France)
ADI	Acceptable Daily Intake
ADP	Adenosine Diphosphate
ADST	Advanced Dual-Sensor Technology
AE	Adverse Event
AEA	Atomic Energy Authority (UK)
AEC	Atomic Energy Commission (US)
AECL	Atomic Energy of Canada Ltd
AF	Anaerobic Filter
AFBC	Atmospheric Fluidized Bed Combustion
AFC	Alkaline Fuel Cell
AFEAS	Alternative Fluorocarbon Environmental Acceptability Study
AFFFBR	Aerobic Fixed Film Fixed Bed Reactor
AFM	Atomic Force Microscope
AGR	Advanced Gas Cooled Reactor
AGU	American Geophysical Union
AHERA	Asbestos Hazard Emergency Response Act (US)
AHU	Air Handling Unit

AIA	Advance Informal Agreement
AIA	Asbestos Information Association
AIChE	American Institute of Chemical Engineers
AID	Agency for International Development (US)
AIHA	American Industrial Hygiene Association
AIHC	American Industrial Health Council
AIMS	Air Information Management System
AIRS	Aerometric Information Retrieval System
AJPH	American Journal of Public Health
AL	Absolute Limen
AL	Action Level
ALAPCO	Association of Local Air Pollution Control Officers (US)
ALARA	As Low As Reasonably Achievable
ALARP	As Low As Reasonably Practicable
ALAT	Alamine Amino Transferase
ALK	Alkalinity
ALLEA	All European Academies
AM	Annual-Maximum
AMA	American Medical Association
AMD	Acid Mine Drainage
AMD	Acrylamide
AMO	Atlantic Multidecadal Oscillation
AMP	Adenosine Monophosphate
AMU	Atomic Mass Unit
ANC	Acid Neutralizing Capacity
ANDRA	National Waste Disposal Organization (France)
ANL	Argonne National Laboratory (US)
ANOVA	Analysis of Variance
ANP	National Petroleum Agency (Brazil)
ANPR	Advanced Notice of Proposed Rulemaking (US)
ANS	American Nuclear Society
ANSI	American National Standards Institute
ANTLE	Affordable Near Term Low Emissions
ANWR	Alaska National Wildlife Refuge
AOAC	Association of Official Agricultural Chemists
AOC	Area of Concern (US)
AOC	Assimilable Organic Carbon
AOCS	American Oil Chemists Society

AOGCM	Atmospheric-Ocean General Circulation Model
AONB	Area of Outstanding Natural Beauty (UK)
AOP	Ammonia Oxidation Process
AOT	Advanced Oxidation Technology
AOX	Organochlorine
APC	Air Pollution Control
APC	Adenomatous Polyposis Coli
APCA	Air Pollution Control Association
APCS	Air Pollution Control System
APDC	Ammonium Pyrolidinodithiocarbonate
APEG	Airborne Particle Expert Group
APEM	Association of Professional Engineers of Manitoba
APHA	American Public Health Association
APHIS	Animal and Plant Health Inspection Service (US)
API	American Petroleum Institute
APS	Aerodynamic Particle Size
APS	American Physical Society
AQCR	Air Quality Control Region (US)
AQEI	Air Quality Exposure Index
AQMD	Air Quality Management District
ARAR	Applicable or Relevant and Appropriate Requirement
ARB	Air Resources Board
ARBRE	Arable Biomass Renewable Energy
ARDS	Acute Respiratory Distress Syndrome
ARM	Aluminized Red Mud
ARM	Availability, Reliability and Maintainability
ARMS	Agricultural Resources Management Study
AROI	Acceptable Range of Oral Intake
ARS	Absorption Refrigeration System
ARS	Agricultural Research Service (US)
AS	Activated Sludge
AS	Australian Standard
ASBR	Anaerobic Sequencing Batch Reactor
ASCE	American Society of Civil Engineers
ASEE	American Society for Engineering Education
ASHRAE	American Society of Heating, Refrigerating and Air Conditioning Engineers
ASLB	Atomic Safety Licensing Board (US)
ASME	American Society of Mechanical Engineers

ASPEN	Advanced Systems for Process Engineering
ASSE	American Society of Safety Engineers
AST	Aboveground Storage Tank
ASTM	American Society for Testing and Materials
ASTMH	American Society for Tropical Medicine and Hygiene
ASTMS	Association of Scientific, Technical and Management Staffs
ASU	Air Separation Unit
ASV	Anode Stripping Voltammetry
ATOC	Acoustic Thermometry of Ocean Climate
ATOFMS	Aerosol Time-of-Flight Mass Spectrometer
ATP	Adenosine Triphosphate
ATP	Advanced Technology Program (US)
ATS	Automatic Tracking Service
ATSDR	Agency for Toxic Substances and Disease Registry (US)
au	Absorbance Unit
AWEP	Association of Women in Environmental Professions
AWG	Association of Women Geoscientists
AWIS	Association for Women in Science
AWMA	Air and Water Management Association
AWRA	American Water Resources Association
AWSE	Association of Women in Science and Engineering (UK)
AWWA	American Water Works Association
B/R	Biomass/Respiration
BABFO	British Association for Biofuels and Oil
BACM	Best Available Control Measure
BACT	Best Available Current (Control) Technology
BAF	Biological Amplification Factor
BALF	Bronchoalveolar Fluid
BANANA	Build Absolutely Nothing ANywhere Anytime
BAP	Benzo-(a)-Pyrene
BAP	Biomass-Associated Product
Bara	Bar Absolute
BART	Best Available Retrofit Technology
BAT	Best Available Technology
BATNEEC	Best Available Technology Not Entailing Excessive Cost (UK)
BBADCP	Broad Band Acoustic Doppler Current Profiler

BBL	Barrels
BCC	Bioaccumulative Chemicals of Concern
BCDL	Beverage Container Deposit Legislation
BCF	Bio-Concentration Factor
BCH	Biosafety Clearing House
BCT	Best Conventional Technology
BD	Butadiene
BDAT	Best Demonstrated Available Technology
BDO	Butanediol
BDOC	Biodegradable Organic Carbon
BDS	Biocatalytic Desulfurization
BEDI	Brownfields Economic Development Initiative
BEP	Board for the Engineering Profession (UK)
BER	Board for Engineering Regulation (UK)
BESS	Battery Energy Storage System
BFR	Brominated Flame Retardant
BFSS	Bioremediation in the Field Search System
BGB	Brilliant Green Bile
BGS	British Geological Survey
BHA	Butylated Hydroxy Anisole
BHC	Benzene Hexachloride
BHC	Lindane
BHET	Bis-Hydroxyethylene Tetraphthalate
BHT	Butylated Hydroxy Toluene
BICER	Baikal International Center for Ecological Research
BID	Background Information Document
BIF	Boilers and Industrial Furnaces
BIOMOVS	Biospheric Model Validation Study
BLM	Bureau of Land Management (US)
BMB	British Medical Bulletin
BMP	Best Management Practices
BN	Base Neutral
BNA	Base Neutral Acid Extractable
BNFL	British Nuclear Fuels plc
BNL	Brookhaven National Laboratory (US)
BNR	Biological Nutrient Removal
BOD	Biochemical Oxygen Demand
BOHC	British Occupational Hygiene Society
BOOS	Burner Out Of Service

BPBR	Batch Packed Bed Reactor
BPD	Biocidal Products Directive (EU)
BPEO	Best Practicable Environmental Option
BPL	Beta Propiolacetone
BPM	Best Practicable Means
BPWTT	Best Practicable Waste Treatment Technology
Bq	Becquerel
BRE	Building Research Establishment (UK)
BREEAM	Building Research Establishment Environmental Assessment Management
BRH	Bureau of Radiological Health (US)
BRWM	Board on Radioactive Waste Management
BS	British Standard
BSE	Bovine Spongiform Encephalopathy
BSERT	Building Services Engineering Research and Technology (UK)
BSI	British Standards Institution
BSJ	Building Services Journal
BSL-4	Biosafety Level 4
BSP	Biomass Support Particle
BSP	Biosafety Protocol
BT	Biotechnology
BTC	Breakthrough Curve
BTEX	Benzene, Toluene, Ethylbenzene and Xylene
BTF	Biotrickling Filter
BTU	British Thermal Unit
BTWC	Biological and Toxin Weapons Convention
BTX	Benzene-Toluene-Xylene
BV	Bed Volume
BWEA	British Wind Energy Association
BWR	Basic Water Requirement
BWR	Boiling Water Reactor
C	Celsius (Centigrade)
C and S	Capture and Storage
C ENG	Chartered Engineer (UK)
CA	Cellulose Acetate
CAA	Clean Air Act (UK) (US)
CAAA	Clean Air Act Amendment (US)
CAD	Computer Aided Design

CAE	Canadian Academy of Engineering
CAE	Computer Aided Engineering
CaFCP	California Fuel Cell Partnership
CAfE	Community Action for Energy
CAFE	Corporate Average Fuel Economy
CAFO	Concentrated Animal Feeding Operations
CAIR	Clean Air Interstate Rule (US)
CAM	Compliance Assurance Monitoring
CAM	Computer Assisted Mapping
CAMP	Continuous Air Monitoring Program
CANDU	Canadian Heavy Water Moderated Reactor
CAP	Chemical Analysis Parameter
CAP	Coagulant Aid Polymer
CAP	Common Agricultural Policy (EU)
CAR	Consolidated Air Rule
CARB	California Air Resources Board
CARL	Conservation and Recreation Lands (US)
CASSIOPEE	Association of Western European Waste Disposal Organizations
CATNIP	Cheapest Available Techniques Not Involving Prosecution (UK)
CATS	Carbon Abatement Technology Strategy
CBA	Cost–Benefit Analysis
CBD	Convention on Biological Diversity
CBED	Convergent Beam Electron Diffraction
CBER	Center for Biologics Evaluation and Research
CBI	Confidential Business Information
CBO	Congressional Budget Office (US)
CBOD	Carbonaceous Biochemical Oxygen Demand
CCAA	Canadian Clean Air Act
CCAMLR	Commission for Conservation of Antarctic Marine Living Resources
CCAR	Closed-Cycle Air Refrigeration
CCCS	Community Climate Change Strategy
CCD	Charged-Coupled-Device
CCDT	Calcium Carbonate Deposition Test
CCGT	Combined Cycle Gas Turbine
CCI	Clean Coal Initiative
CCL	Climate Change Levy

CCL	Contaminant Candidate List (US)
CCME	Canadian Council of Ministers of the Environment
CCMS	Committee on the Challenges of Modern Society
CCOL	Coordinating Committee on the Ozone Layer
CCP	Cities for Climate Protection
CCP	Combined Cool and Power
CCPE	Canadian Council of Professional Engineers
CCR	Council for Chemical Research
CCS	Carbon Dioxide Capture and Storage
CCSM	Community Climate System Model
CCT	Clean Coal Technology
CDAA .	2-Chloro-N, N-Diallyl Acetamide
CDC	Centers for Disease Control and Prevention (US)
CDC	Communicable Disease Center (US)
CDEC	2-Chlorallyl-N, N-Diethyldithiocarbamate
CDER	Center for Drug Evaluation and Research
CDI	Capacitive Deionization
CDI	Continuous Deionization
CDI	Critical Density Index
CDM	Clean Development Mechanism
CDOM	Chromophoric Dissolved Organic Material
CDRH	Center for Devices and Radiological Health
CDWG	Canadian Drinking Water Guidelines
CEA	Cumulative Environmental Assessment
CEA	Atomic Energy Commission (France)
CEAA	Canadian Environmental Assessment Act
CEAB	Canadian Engineering Accreditation Board
CEARC	Canadian Environmental Assessment Research Council
CEC	Cation Exchange Capacity
CEC	Commission of the European Communities
CECAB	Canadian Environmental Certification Approvals Board
CEDI	Continuous Electrode-Deionization
CEES	Complex Ecological-Economic System
Cefic	European Chemical Industry Council
CEGB	Central Electricity Generating Board (UK)
CEM	Continuous Emissions Monitoring
CEMS	Continuous Emission Monitoring System
CEN	European Committee for Standardization
CEN/SCK	Nuclear Research Organization (Belgium)

CEPA	Canadian Environmental Protection Act
CEQ	Council on Environmental Quality (US)
CER	Certified Emission Reduction
CERCLA	Comprehensive Environmental Response, Compensation and Liability Act (US)CERN
	European Laboratory for Particle Physics
CESER	Countermeasures: Environmental and Socio-Economic Responses
CETP	Common Effluent Treatment Plant
CFB	Circulating Fluidized Bed
CFBC	Circulating Fluidized Bed Combustion
CFBC	Circulating Fluidized Bed Combustor
CFC	Chlorofluorocarbons
CFC	1,1,2-Trichloro-1,2,2-Trifluoroethane
CFD	Computational Fluid Dynamics
CFM	Cubic Feet per Minute
CFR	Code of Federal Regulations (US)
CFR	Continuous Flow Reactor
CFS	Canadian Forest Service
CFS	Cubic Feet per Second
CFSTR	Continuous Flow Stirred Tank Reactor
CFV	Clean Fuel Vehicle
CG/HCCS	Coordinating Group for the Harmonization of Chemical Classification Systems
CGCM	Coupled General Circulation Model
CGDF	Coal Gasification Development Facility
CGIAR	Cooperative Group of International Agricultural Research Centers
CGL	Comprehensive General Liability
CGMP	Current Good Manufacturing Practice
CGSB	Canadian General Standards Board
CHIP	Chemical Hazards Information Profile
CHL	Chlorophyll
CHL a	Chlorophyll a
CHONSP	Carbon, Hydrogen, Oxygen, Nitrogen, Sulfur, Phosphorous
CHP	Combined Heat and Power
CHP/HP	Combined Heat and Power/Heat Pump
CHPA	Combined Heat and Power Association

CHS	Contact Hypersensitivity
CI	Confidence Interval
CI	Cumulative Impact
Ci	Curie
CIA	Chemical Industries Association (UK)
CIA	Cumulative Impact Assessment
CIB	Conseil International du Baitment Pour Recherche, l'Etude et la Documentation
CIBSE	Chartered Institution of Building Services Engineers (UK)
CIEH	Chartered Institution of Environmental Health (UK)
CIH	Chartered Institute of Housing (UK)
CIP	Clean In Place
CISTI	Canadian Institute for Science and Technical Information
CITES	Convention on International Traffic in Endangered Species
CJD	Creutzfeldt-Jacob Disease
CL	Criterion Level
CLAIRE	Contaminated Land Applications in Real Environments
CLARINET	Contaminated Land Rehabilitation Networks for Environmental Technologies
CLD	Chemiluminescence Detector
CLER	Carbon-Limited Enrichment
CLIMAP	Climate-Leaf Analysis Multivariate Program
CLIMAPP	Climate/Long Range Investigation Mapping and Prediction Project
CM	Combustible Matter
CMA	Chemical Manufacturers Association (US)
CMC	Critical Micelle Concentration
CMD	Count Mean Diameter
CMDL	Climate Monitoring and Diagnostics Laboratory (US)
CMPU	Chemical Manufacturing Process Unit
CMR	Complete Mix Reactor
CNC	Condensation Nucleus Counter
CNG	Compressed Natural Gas
CNI	Consiglio Nazionale Ingegneri (Italy)
CNISF	Conseil National des Ingenieurs et Scientifiques de France
CNL	Compound Noise Level

CNL	Corrected Noise Level
CNP	National Research Council (Italy)
CNRS	Centre National de la Recherche Scientifique (France)
CNSIF	National Council of French Engineers and Scientists
CNT	Carbon Nanotube
CO_2eq	Carbon Dioxide Equivalent
COD	Chemical Oxygen Demand
COGEMA	Nuclear Waste Processing Organization (France)
COH	Coefficient of Haze
COHb	Carboxyhemoglobin
COM	Coastal Oceanographic Model
COM	Continuous Opacity Monitor
COMEAP	Committee on the Medical Effects of Air Pollution
CONCAWE	Conservation and Clean Air and Water in Europe
COP	Coefficient of Performance
COP	Conference of the (Contracting) Parties (UNFCCC)
COP/MOP2	Conference of the Parties Serving as the Meeting of the Meeting of the Parties to the Kyoto Protocol
CoPA	Control of Pollution Act (UK)
Co-PCB	Coplanar Polychlorinated Biphenyl
COPD	Chronic Obstructive Pulmonary Disease
COPE	Committee On Publication Ethics
COPUS	Committee on the Public Understanding of Science (UK)
CoRWM	Committee on Radioactive Waste Management (UK)
COS	Carbonyl Sulfide
COVRA	Nuclear Waste Disposal Organization (Holland)
CP	Coarse Particle
CPC	Condensation Particle Counter
CPCB	Central Pollution Control Board (India)
CPF	Cancer Potency Factor
CPI	Chemical Process Industries
Cpm	Counts per Minute
cps	Counts per Second
CPSC	Consumer Product Safety Commission (US)
CPVC	Chlorinated Polyvinyl Chloride
CQC	2,6 Dichloroquinonechlorimide
CR	Community Respiration
CR	Conditioned Response

CRADA	Cooperative Research and Development Agreement
CREATE	Centre for Research, Education and Training in Energy
CRNL	Chalk River Nuclear Laboratories (Canada)
CRT	Continuously Regenerating Trap
CS	Conditioned Stimulus
CSB	Chemical Safety (and Hazards Investigation) Board
CSCE	Canadian Society for Civil Engineering
CSD	Commission on Sustainable Development
CSH	Calcium Silicate Hydrate
CSI	Common Sense Initiative (US)
CSIR	Council for Scientific and Industrial Research (India) (South Africa)
CSIRO	Commonwealth Scientific and Industrial Research Organization (Australia)
CSMA	Chemical Specialties Manufacturers Association
CSO	Combined Sewer Overflow
CSOD	Carbonaceous Sediment Oxygen Demand
CSPE	Canadian Society for Professional Engineers
CSPI	Center for Science in the Public Interest
CSR	Center for Scientific Review (US)
CSR	Corporate Social Responsibility
CSS	Combined Sewer System
CSSP	Council of Scientific Society Presidents
CST	Capillary Suction Time
CST	Carbon Storage Trust
CSTR	Continuous Flow Stirred Tank Reactor
CT	Carbon Tetrachloride
CTAB	Cetyltrimethyl ammonium bromide
CTG	Control Techniques Guideline
CTI	Climate Technology Initiative
CTR	Council for Tobacco Research
CTS	Compound Threshold Shift
CU	Common Usage
CUSEC	Cubic Feet per Second
CUTE	Clean Urban Transport for Europe
CV	Calorific Value
CVI	Children's Vaccine Initiative
CVS	Constant Volume Sampler
CWA	Chemical Warfare Agent

CWA	Clean Water Act (US)
CWC	Chemical Weapons Convention
CWO	Catalytic Wet Oxidation
CWS	Coal–Water Slurry
CZARA	Coastal Zone Act Reauthorization Amendments
CZMA	Coastal Zone Management Act (US)
D/DBP	Disinfectant/Disinfection Byproduct Rule (US)
D/F	Dioxin/Furan
DAD	Decide, Announce and Defend
DADA	Decide, Announce, Defend and Abandon
DAF	Dissolved Air Flotation
DALY	Disability Adjusted Life Years
DAS	Data Acquisition System
dB	Decibel
dB(A)	Decibel A–Weighted
DBMCA	Dibromo-p-methyl-carboxyazo
DBMS	Database Management Systems
DBOOM	Design, Build, Own, Operate and Maintain
DBP	Disinfection By-Product
DBPC	Di Tertiary Butyl-P-Cresol
DBPP	Disinfection By-Product Precursor
DBT	Dibutyltin
DBT	Dimethyldibenzothiophene
DC	Delta Commission (Holland)
DCB	4,4′-Dichlorobiphenyl
DCE	1,2-Dichloroethylene
DCER	Dwelling Carbon Emission Rate
DCL	Design Control Limit
DCL	Direct Coal Liquefaction
DD	Dynamic Demand
DDA	4,4′-Dichlorodiphenylacetic Acid
DDD	1,1-Dichloro-2,2-Bis(p-chlorophenyl)-Ethane
DDE	1,1-Dichloro-2,2-Bis(p-chlorophenyl)-Ethylene
DDT	Dichlorodiphenyltrichloroethane
DDVP	2,2-Dichlorovinyl Dimethyl Phosphate
DE	Diatomaceous Earth
DEA	Diethanolamine
DEC	Department of Environmental Conservation
DEE	1,2-Diethoxyethane

DEET	Diethyltoluamide
DEFRA	Department of the Environment, Food and Rural Affairs (UK)
DEG	Diethylene Glycol
DEP	Department of Environmental Protection
DER	Discrete Emission Reduction (US)
DERC	Discrete Emission Reduction Credits (US)
DES	Diethylstilbesterol
DETR	Department of the Environment, Transport and Regions (UK)
DFB	Diffusion Fiber Bed
DfT	Department for Transportation (UK)
DHA	District Health Authority (UK)
DHEP	Di(2-Ethylhexyl) Phthalate
DHEW	Department of Health, Education and Welfare (US)
DHHS	Department of Health and Human Services (US)
DHSS	Department of Health and Social Security (UK)
DIAL	Differential Absorption Lidar
DIH	Division of Industrial Hygiene (US)
DiP	Discussion in Principal
DISH	Deepwater Installation of Subsea Hardware
DL	Detection Limit
DL	Difference Limen
DMA	Differential Mobility Analyzer
DMDS	Dimethyldisulfide
DME	Dimethyl Ether
DMHRF	Dual Media High Rate Filtration
DMI	Danish Meteorological Institute
DMR	Discharge Monitoring Report
DMS	Dimethyl sulfide
DMSO	Dimethyl sulfoxide
DMT	Dimethyl Tetraphthalate
DNA	Deoxyribonucleic Acid
DNAPL	Dense Non-Aqueous Phase Liquid
DNB	Dinitrobenzene
DNIP	Di-Isononyl Phthalate
DO	Dissolved Oxygen
DOC	Dissolved Organic Carbon
DOD	Department of Defense (US)

DOE	Department of Energy (UK) (US)
DoE	Department of the Environment (UK)
DoH	Department of Health (UK)
DOJ	Department of Justice (US)
DOM	Dissolved Organic Matter
DOT	Department of Transport(ation) (Canada) (UK) (US)
DPBA	Diphenylbenzamidine
DPE	1,1-Diphenylethylene
DPEI	Department of Productivity, Energy and Industry (UK)
DPKO	Dipyridyl Ketone Oxime
dpm	Disintegrations Per Minute
dps	Disintegrations Per Second
DPW	Department of Public Works
DQI	Data Quality Indicator
DQO	Data Quality Objective
DRE	Destruction and Removal Efficiency
DRI	Development of Regional Impact
DRM	Dynamic River Management
DS	Degree of Succession
DS	Dry Scrubber
DS	Dry Solids
DSCD	Danish Research Agency Committee on Scientific Dishonesty
DSDP	Deep Sea Drilling Project
DSM	Demand Side Management
DSR	Dissimilatory Sulfate Reducing
DTH	Delayed Type Hypersensitivity
DTI	Department of Trade and Industry (UK)
DTLR	Department for Transport, Local Government and the Regions (UK)
DTPA	Diethylenetriaminepentaacetic Acid
DWF	Dry Weather Flow
DWL	Derived Working Limit
DWT	Deadweight Ton
E and C	Engineering and Construction
EA	Energy Amplifier
EA	Environmental (Agency) (Authority) (UK)
EA	Environmental Assessment
EA	Environmental Audit

EAC	Extruded Activated Carbon
EAEC	European Atomic Energy Community (EU)
EAGE	European Association of Geoscientists and Engineers
EAGGF	European Agriculture Guidance and Guarantee Fund
EANET	Acid Deposition Monitoring Network in East Asia
EAP	Emergency Action Plan
EAP	Environment Action Program (EU)
EAPS	Environmental Aspects in Product Standards
EARP	Environmental Assessment and Review Process
EASST	European Association for the Study of Science and Technology
EB	Ethylene Glycol Monobutyl Ether
EB/S	Ethylbenzene/Styrene
EBDC	Ethylene Bisdithiocarbamate
EBI	European Bioinformatics Institute
EBRD	European Bank of Reconstruction and Development
EC	Electrical Conductivity
EC	Elemental Carbon
EC	Engineering Council (UK)
EC	European Code
EC	European Commission
EC	European Community
EC2000	Engineering Criteria 2000
EC_{50}	Mean Effective Concentration
ECBM	Enhanced Coalbed Methane Recovery
ECCP	European Climate Change Programme
ECDIN	Environmental Chemical Data Information Network (EU)
ECE	UN Economic Commission for Europe
ECEPA	Environmental Challenge for European Port Authorities
ECF	Elemental Chlorine Free
ECHS	Environmental Clearing House System
ECL	Exposure Control Limit
ECLSS	Environmental Control and Life Support System
ECMWF	European Center For Medium Range Weather Forecasting
ECOSOC	Economic and Social Council
ECTFE	European Fluorochemicals Technical Committee
ECU	Extractable Copper

ED	Electrodialysis
ED	Exposure Duration
ED	Extractive Distillation
EDC	Electron Donor Compound
EDC	Ethylenedichloride
EDC	European Documentation Centre
EDF	Environmental Defense Fund
EDF	European Development Fund
EDI	Electrodeionization
EDI	Electronic Data Interchange
EDIP	Environmental Design of Industrial Products
EDR	Electrodialysis Removal
EDR	Electrodialysis Reversal
EDTA	Ethylenediaminetetraacetate
EDXRA	Energy Dispersive X-Ray Analysis
EEA	Economic Espionage Act (US)
EEA	European Environment Agency
EEB	European Environment Bureau
EEBPP	Energy Efficiency Best Practices Program
EEC	European Economic Community
EELS	Electron Energy Loss Spectroscopy
EEMD	Environmental and Energy Management Directorate (UK)
EERE	Energy Efficiency and Renewable Energy
EF	Emissions Factor
EFCE	European Federation of Chemical Engineers
EFCEA	European Federation of Consulting Engineers Associations
EFFS	European Flood Forecasting System
EFR	External Floating Roof
EfW	Energy from Waste
EG	Ethylene Glycol
EGA	Exhaust Gas Analyzer
EGL	Energy Grade Line
EGR	Enhanced Gas Recovery
EGR	Exhaust Gas Recirculation
EGSB	Expanded Granular Sludge Bed
EGTC	Exhaust Gas Test Cell
EHC	Environmental Health Criteria

EHO	Environmental Health Officer (UK)
EHP	Environmental Health Perspectives
EHS	Environmental Health and Safety
EHS	Extremely Hazardous Substance
EI	Energy Institute (UK)
EIA	Energy Information Administration (US)
EIA	Environmental Impact Assessment
EIA	Environmental Investigation Agency (UK)
EIB	European Investment Bank
EIC	Energy Industry Council
EIC	Engineering Institute of Canada
EINECS	European Inventory of Existing Commercial Chemical Substances (EU)
EIS	Environmental Impact Statement
EJ	Environmental Justice
EKMA	Empirical Kinetic Modeling Approach
ELC	Environmental Liaison Center
ELNES	Energy Loss Near Edge Structure
ELR	Environmental Lapse rate
eM&T	Energy Monitoring and Targeting
EMAS	Ecoaudit and Management Scheme (EU)
EMAS	Employment Medical Advisory Service (UK)
EMB	Eosin Methylene Blue
EMC	Event Mean Concentration
EMEA	European Agency for Evaluation of Medicinal Products
EMEC	European Marine Energy Centre
EMEP	European Monitoring and Evaluation Programme
EMF	Electromagnetic Field
EMIS	Environmental Management Information System
EML	Environmental Measurements Laboratory (US)
EMP	Emergency Management Plan
EMR	Electromagnetic Radiation
EMS	Environmental Management Systems
EMTIC	Emission Measurement Technical Information Center (US)
EN	European Committee for Standardization
ENEA	Italian National Agency for New Technology, Energy and the Environment
EngC	Engineering Council (UK)

ENHPA	European Network of Health Protection Agencies
ENSO	El Nino-Southern Oscillation
EO	Ethylene Oxide
EO	Executive Order (US)
EOR	Enhanced Oil Recovery
EOS	Earth Observation (Observing) System
EPA	Energy Policy Act (US)
EPA	Environmental Protection Act (UK)
EPA	Environmental Protection Agency (US)
EPB	Extractable Lead
EPCRA	Emergency Planning and Community Right-To-Know Act (US)
EPE	Environmental Performance Evaluation
EPEC	Enteropathogen E. coli
EPER	European Pollutant Emissions Register
EPFT	Swiss Federal Institute of Technology
EPH	Extractable Petroleum Hydrocarbons
EPHA	European Public Health Alliance
EPM	Environmental Pressure Measurement
EPMA	Electron Probe Micro Analysis
EPNdB	Effective Perceived Noise Level
EPR	European Pressurized Reactor
EPR	Extended Producer Responsibility
EPS	Extracellular Polymeric Substance
EPS	Engineering Process Safety
EPSRC	Engineering and Physical Sciences Research Council (UK)
EPTC	S-ethyl,N,N,-dipropylthiocarbamate
eq	Equivalent
eq/L	Equivalents per Liter
EQIP	Environmental Quality Incentive Program
EQO	Environmental Quality Objective
ER	Environmental Review
ERA	Environmental Risk Assessment
ERC	Emissions Reductions Credits
ERDA	Energy Research and Development Administration (US)
ERDF	European Regional Development Fund
ERP	Emergency Response Plan

ERP	Enterprise Resource Planning
ERS	Environmental Remote Sensing
ERTS	Earth Resources Technology Satellite
ERU	Emission Reduction Unit
ESA	Ecological Society of America
ESA	Endangered Species Act (US)
ESD	Electrostatic Dissipation
ESEF	European Science and Environment Forum
ESEM	Earth Systems Engineering and Management
ESEM	Environmental Scanning Electron Microscope
ESEP	European Science Exchange Program
ESF	Exploratory Studies Facility
ESF	European Science Foundation
ESP	Electrostatic Precipitator
ESR	Electron Spin Resonance
ESR	Environmental Screening Review
EST	Energy Saving Trust (UK)
ESV	Equivalent Sample Volume
ETA	Ecologically Threatened Area
ETAN	European Technology Assessment Network
ETB	Engineering and Technology Board (UK)
ETBE	Ethyl Tert-Butyl Ether
ETC/S	European Topic Center on Soil
ETG	Emissions Trading Group
ETH	Swiss Federal Institute of Technology
ETO	Ethylene Oxide
ETR	Ecological Tax Reform (UK)
ETS	Emissions Tracking System
ETS	Emissions Trading Scheme
ETS	Environmental Tobacco Smoke
ETSU	Energy Technology Support Unit (UK)
ETT	Effluent Toxicity Treatability
ETU	Ethylene Thiourea
EU	European Union
EU EST	European Union Energy Saving Trust
EUR ING	European Engineer
EURADWASTE	Nuclear Waste Reports (CEC)
Euratom	European Atomic Energy
EUROMETAUX	European Association of Metals

EuroREX	European Renewable Energy Exchange
EV	Electric Vehicle
eV	Electron Volt
EWEA	European Wind Energy Association
EWG	Environmental Working Group
EZN	Extractable Zinc
F	Fahrenheit
f/cc	Fibers per Cubic Centimeter
FACA	Federal Advisory Committee Act (US)
FAIR	Federal Agriculture Improvement and Reform Act (US)
FAME	Fatty Acid Methyl Ester
FAO	Food and Agriculture Organization (UN)
FAP	Filter Aid Polymer
FAR	Fatal Accident Rate
FAZ	Fly Ash Based Zeolite
FBC	Fluidized Bed Combustion
FBI	Family Biotic Index
FC	Fecal Coliforms
FC	Fuel Cell
FCC	Fluid Catalytic Cracker
FCCC	Framework Convention on Climate Change
FCM	Fuel Containing Mass
FCO	Foreign and Commonwealth Office (UK)
FD	Formamide
FDA	Food and Drug Administration (US)
FDD	Fault Detection and Diagnostics
FEA	Finite Element Analysis
FEANI	European Federation of National Engineering Associations
FEG	Field Emission Electron Gun
FEH	Flood Estimation Handbook
FEI	Fire and Explosion Index
FEMA	Federal Emergency Management Agency (US)
FERC	Federal Energy Regulatory Commission (US)
FETC	Federal Energy Technology Center (US)
FEV	Forced Expired Volume
FF	Fabric Filter
FF	Future Forests
FFA	Federal Facilities Agreement (US)
FFA	Free Fatty Acid

FFCA	Federal Facilities Compliance Act (US)
FFDCA	Federal Food, Drug and Cosmetic Act (US)
FFF	Forum for the Future
FGD	Flue Gas Desulfurization
FGR	Flue Gas Recirculation
FHSA	Family Health Service Authority (UK)
FIA	Flame Ionization Analyzer
FIA	Flow Injection Analysis
FID	Flame Ionization Detector
FIFRA	Federal Insecticide, Fungicide and Rodenticide Act (US)
FIRM	Flood Insurance Rate Map
FISEC	Foundation for Scientific and Educational Cooperation
FM	Fatty Matter
FML	Flexible Membrane Liner
FoE	Friends of the Earth
FONSI	Finding of No Significant Impact
FP	Fine Particle
FP6	Sixth Framework Program (EU)
FPC	Federal Power Commission (US)
FPD	Flame Photometric Detector
FQPA	Food Quality Protection Act (US)
FR	Federal Register (US)
FRC	Functional Residual Capacity
FRP	Fiberglass Reinforced Plastics
FS	Fecal Streptococci
FSA	Food Standards Agency (UK)
FSD	Full Scale Demonstration
FSR	Final Safety Report
FSU	Former Soviet Union
FT	Fischer–Tropsch
FT	Fourier Transform
FTA	Fault Tree Analysis
FTC	Federal Trade Commission (US)
FTO	Flameless Thermal Oxidation
FVC	Forced Vital Capacity
FWPCA	Federal Water Pollution Control Act (US)
FWQA	Federal Water Quality Administration (US)
FWS	Fish and Wildlife Service (US)

FWS	Free Water Surface
FY	Fiscal Year
g.l.c.	Ground Level Concentration
G7	Group of Seven
G8	Group of Eight
GAC	Granular Activated Carbon
GAMOT	Global Acoustic Mapping of Ocean Temperatures
GAO	General Accounting Office (US)
GATS	General Agreement on Tariffs and Services
GATT	General Agreement on Trade and Tariffs
Gb	Giga Barrels
GC	Gas Chromatograph
GCCI	Global Climate Change Initiative
GCM	General Circulation Model
GCM	Global-Circulation Model
GC-MS	Gas Chromatography and Mass Spectrometry
GCP	Good Combustion Practice
GDI	Gasoline Direct Injection
GDP	Gross Domestic Product
GEF	Global Environment Facility
GEM	Genetically Engineered Microbe
GEMI	Global Environmental Management Initiative
GEP	Global Environmental Program
GEV	Generalized Extreme Value
GEV	Gumbel Extreme Value
GEWEX	Global Energy and Water Cycle Experiment
GFDL	Geophysical Fluid Dynamics Laboratory
$gft^{-2}d^{-1}$	Gallons per Square Foot per Day
GHG	Greenhouse Gas
GHS	Globally Harmonized System
GIS	Geographic Information Systems
GISS	Goddard Institute of Space Studies
GJ	Giga Joules
GLIER	Great Lakes Institute for Environment Research
GLM	Generalized Linear Model
GLP	Good Laboratory Practice
GLUE	Generalized Likelihood Uncertainty Estimation
GLUMRB	Great Lakes-Upper Mississippi River Board (US)
GM	Genetically Modified

GMA	General Mining Act (US)
GMO	Genetically Modified Organism
GNP	Gross National Product
GOR	Gain Output Ratio
gpcd	Gallons Per Capita Per Day
GPD	Gallons Per Day
GPH	Gallons Per Hour
GPM	Gallons Per Minute
GPO	Government Printing Office (US)
GPP	Gross Primary Production
GPS	Global Positioning System
GRACE	Gravity Recovery and Climate Experiment
GRAS	Generally Recognized As Safe
GRAS/E	Generally Recognized As Safe and Effective
GRI	Gas Research Institute (US)
GRIP	Greenland Ice Core Project
GRR	Gross Reproduction Rate
GSA	Geothermal Steam Act (US)
GSF	Nuclear Research Organization (Germany)
GSH-Px	Glutathione Peroxidase
GtC	Gigatons Carbon
GTL	Gas to Liquid
GTN	Global Trend Network
GV	Guideline Value
GVF	Gas Void Fraction
GVWHO	World Health Organization Guideline Value
GWh	Gigawatt Hour
GWP	Global Warming Potential
Gy	Gray
H6CB	3,3',4,4',5-Hexachlorinated Biphenyl
ha	Hectare
HA	Humic Acid
HAA	Haloacetic Acids
HAc	Acetic Acid
HACCP	Hazard Analysis of Critical Control Point
HAD	Hazards Assessment Document
HAP	Hazardous Air Pollutant
HAZMAT	Hazardous Material
HAZOP	Hazard and Operability Study

HBD	2-Hydroxyisobutyl Amide
HBM	2-Hydroxyisobutric Acid Methyl Ester
HC	Hydrocarbon
HCFC	Hydrochlorofluorocarbon
HCS	Hazard Communication Standard
HDM	Home Dust Mite
HDPE	High Density Polyethylene
HDS	Hydrodesulfurization
HDTMABr	Hexadecyltrimethyl Ammonium Bromide
HECA	Home Energy Conservation Act (UK)
HEES	Home Energy Efficiency Scheme (UK)
HEI	Health Effects Institute
HEL	Highly Erodible Land
HELCOM	Helsinki Commission
HEPA	High Efficiency Particulate Air
HERP	Human Exposure/Rodent Potency
HES	Human Embryonic Stem
HEST	Hazardous Element Sampling Train
HEV	Hybrid Electric Vehicle
HEW	Department of Health, Education and Welfare (US)
HFC	Hydrofluorocarbon
HFE	Hydrofluoroether
HFO	Heavy Fuel Oil
HGH	Human Growth Hormone
HGL	Hydraulic Grade Line
HGM	Hazardous Gas Monitor
HGMS	High Gradient Magnetic Separation
HHS	Department of Health and Human Services (US)
HIPPS	High Performance Power Systems
HIUI	Health Institute of the Uranium Industry (Czechoslovakia)
HIV	Human Immuno-Deficiency
HLB	Hydrophilic-Lipophilic Balance
HLLW	High Level Liquid Waste
HLW	High Level Radioactive Waste
HM	Heavy Metal
HMDE	Hanging Mercury Drop Electrode
HMIP	HM Inspectorate of Pollution (UK)
HMNII	HM Nuclear Installations Inspectorate (UK)
HMSO	HM Stationery Office

HON	Hazardous Organic Neshap (National Emissions Standards for Hazardous Air Pollutants) (US)
HPA	Health Protection Agency (UK)
HPA	Hetero Polyanion
HPC	Heterotrophic Plate Count
HPLC	High Performance Liquid Chromatography
HPS	Health Physics Society
HQ	Health Quotient
HR	Humidity Ratio
HRA	Halogen Resistant Azole
HRGC	High Resolution Gas Chromatography
HRS	Hazard Ranking Score (US)
HRSA	Health Resources and Services Administration (US)
HRT	Hydraulic Residence Time
HRTEM	High-Resolution Transmission Electron Microscopy
HSC	Health and Safety Commission (UK)
HSCT	High Speed Civil Transport
HSDB	Hazardous Substances Data Base
HSE	Health and Safety Executive (UK)
HSWA	Hazardous and Solid Waste Amendments (US)
HSWA	Health and Safety at Work Act (UK)
HTE	Horizontal Tube Evaporator
HTF	Heat Transfer Fluid
HTFW	High Temperature Fluid Wall
HTH	High Test Hypochlorite
HTlc	Hydrotalcite-Like-Compounds
HTME	Horizontal Tube Multiple Evaporator
HTRW	Hazardous, Toxic and Radioactive Waste
HTU	Height of Transfer Unit
HTU	Hydro Thermal Upgrading
HUD	Housing and Urban Development
HVAC	Heating, Ventilating and Air Conditioning
HVOC	Halogenated Volatile Organic Compound
HW	Hydrological Weir
HWC	Hazardous Waste Combustion
HWE	Healthy Worker Effect
HWI	Hazardous Wastes Inspectorate (UK)
HWIR	Hazardous Waste Identification Rule (US)
HWL	High Water Level

Hz	Hertz
I/A	Innovative and Alternative
I/I	Infiltration/Inflow
I/O	Input/Output
IAEA	International Atomic Energy Agency
IAHEE	International Association for Hydrogen Energy
IAP	Indoor Air Pollution
IAP	InterAcademy Panel
IAQ	Indoor Air Quality
IARC	International Agency for Research on Cancer
IAWPRC	International Association on Water Pollution Research and Control
IAWQ	International Association on Water Quality
IBC	International Bulk Container
IBRD	International Bank for Reconstruction and Development (World Bank)
IC	Internal Combustion
IC_{50}	Mean Inhibitory Concentration
ICAA	International Council of Chemical Associations
ICAR	Indian Council of Agricultural Research
ICBEN	International Commission on the Biological Effects of Noise
ICBG	International Cooperative Biodiversity Grants Program
ICCP	Intergovernmental Committee for the Cartagena Protocol
ICD	International List of Diseases, Injuries and Causes of Death
ICE	Institution of Civil Engineers (UK)
ICES	International Council For Exploration of the Seas
ICHARM	International Centre For Water Hazard and Risk Management (Japan)
IChE	Institution of Chemical Engineers
ICIDI	Independent Commission on International Development Issues
ICL	Indirect Coal Liquefaction
ICP	Integrated Contingency Plan (US)
ICP	International Cooperative Program
ICPD	International Conference on Population and Development

ICP-OES	Inductively Coupled Plasma Optical Emission
ICR	Industrial Cost Recovery
ICR	Information Correction Rule
ICRAM	Central Institute for Scientific and Technological Research Applied to the Sea (Italy)
ICRP	International Commission on Radiological Protection
ICSEAF	International Commission for Southeast Atlantic Fisheries
ICSU	International Council of Scientific Unions
ICT	Information Communications Technology
ICT	Intercontinental Transport
ICT	International Critical Table
ICZM	Integrated Coastal Zone Management
IDA	International Development Association
IDAA	International Development Assistance Agency
IDEA	International Development Enterprise Associates
IDF	Intensity Duration Frequency
IDGEC	Institutional Dimensions of Global Environmental Change
IDLH	Immediately Dangerous to Life and Health
IDNDR	International Decade for Natural Disaster Reduction
IDSE	Initial Distribution System Evaluation
IDWSSD	International Drinking Water Supply and Sanitation Decade
IE	Cationic Exchange
IE	Institute of Energy (UK)
IEA	International Energy Agency
IEC	International Electrotechnical Commission
IEE	Institute of Electrical Engineers (UK)
IEEE	Institute of Electrical and Electronics Engineers
IESWTR	Interim Enhanced Surface Water Treatment Rule (US)
IFB	Impaction Fiber Bed
IFC	International Finance Corporation
IFGR	Induced Flue Gas Recirculation
IFP	Institut Francais du Petrole (France)
IFR	Internal Floating Roof
IGBP	International Geosphere-Biosphere Program
IGC	Index of Grassland Condition
IGCC	Integrated Gasification Combined Cycle
IGO	Intergovernmental Organization
IGS	Inert Gas System

IGT	Institute of Gas Technology
IGY	International Geophysical Year
IHS	Indian Health Service (US)
IIASA	International Institute for Applied Systems Analysis
IIEC	International Institute for Energy Conservation
IIHEE	International Institute for Hydraulic and Environmental Engineering
IJC	International Joint Commission
IKSR	International Commission for the Protection of the Rhine Against Pollution
ILGRA	Interdepartmental Liaison Group on Risk Assessment (UK)
ILO	International Labor Organization
ILW	Intermediate Level Radioactive Waste
IM(ech)E	Institution of Mechanical Engineers (UK)
IMDG	International Maritime Dangerous Goods Code
IMF	International Monetary Fund
IMI	Imidazolinone
IMIQ	Mexican Institute of Chemical Engineers
IMO	International Maritime Organization
IMR	Infant Mortality Rate
IMS-ICS	Interagency Incident Management System-Incident Command System (US)
INEEL	Idaho National Engineering and Environmental Laboratory (US)
INSA	Institut National des Sciences Appliques (France)
InstE	Institute of Energy
INTRAVAL	International Geosphere Transport Model Validation
INWAC	IAEA Advisory Committee
IOC	Inorganic Chemical
IOD	Immediate Oxygen Demand
IoE	Institute of Energy
IOH	Institute of Occupational Hygiene (UK)
IOM	Institute of Medicine (US)
IoP	Institute of Physics (UK)
IOP	Indian Ocean Dipole
IOPC	International Oil Pollution Compensation Fund
IOPP	International Oil Pollution Prevention (Certificate)
IOWG	International Orimulsion Working Group

IP	Inhalable Particles
IP	Institute of Petroleum
IPA	Isopropanol
IPACT	International Pharmaceutical Aerosol Consortium for Toxicology Testing
IPAT	Impact, Population, Affluence, Technology
IPC	Integrated Pollution Control (UK)
IPCC	Intergovernmental Panel on Climate Change
IPCS	International Program on Chemical Safety
IPI	International Population Institutions
IPM	Inhalable Particulate Matter
IPM	Integrated Pest Management
IPOD	International Programme of Ocean Drilling
IPP	Independent Power Producer
IPPC	Integrated Pollution Prevention and Control (EC)
IPPR	Institute for Public Policy Research
IPTS	Institute for Prospective Technological Studies
IR	Inactivation Ratio
IR	Infrared
IRAA	Indoor Radon Abatement Act (US)
IRB	Institutional Review Board
IRIS	Integrated Risk Information System (US)
IRP	Inward Processing Release
IRPP	Institute for Research on Public Policy
IRPTC	International Register of Potentially Toxic Chemicals
IRR	Ionizing Radiations Regulations (UK)
IRUS	Intrusion Resistant Underground Structures
ISA	Ideologically Structured Action
ISA	International Seabed Authority
ISC	Interstate Sanitation Commission
ISES	International Solar Energy Society
ISI	Indian Standards Institution
ISM	Industrial, Scientific, Medical
ISO	Independent System Operator
ISO	International Standards Organization
ISPE	International Society of Pharmaceutical Engineers
ISRM	In-Situ Redox Manipulation
ISSX	International Society for the Study of Xenobiotics
ISTAS	International Symposium on Technology and Society

IT	Information Technology
ITC	Interagency Testing Committee (US)
ITCZ	Inter-Tropical Convergence Zone
ITD	Ion Trap Detector
ITF	Industry Technology Facilitator
ITM	Ionic Transport Membrane
IUCN	International Union for Conservation of Nature and Natural Resources
IUFoST	International Union of Food Science and Technology
IUGG	International Union of Geodesy and Geophysics
IUPAC	International Union of Pure and Applied Chemistry
IURE	Inhalation Unit Risk Estimate
IWC	International Whaling Commission
IWM	Integrated Waste Management
IWPCC	Interstate Water Pollution Control Compact (US)
IWS	Ionizing Wet Scrubber
IX	Ion Exchange
J	Joule
JAEC	Japanese Atomic Energy Commission
JAERI	Japanese Atomic Energy Research Institute
JAPCA	Journal of the Air Pollution Control Association
JAWWA	Journal of the American Water Works Association
JBR	Jet Bubbling Reactor
JECFA	Joint Expert Committee on Food Additives
JGOFS	Joint Global Ocean Flux Study
JI	Joint Implementation
JI	Joint Investment
JMPR	Joint Meeting on Pesticide Residues
JRC	Joint Research Center
JRSH	Journal of the Royal Society for the Promotion of Health
JTU	Jackson Turbidity Unit
JWPCF	Journal of the Water Pollution Control Federation
K	Kelvin
KASAM	National Council for Nuclear Waste (Sweden)
KAST	Korean Academy of Science and Technology
Kb	Kilobar
$kgm^{-3}d^{-1}$	Kilogram per cubic meters per day
KJ	Kilojoule

Kn	Knudsen Number
KT	Clearness Index
kV	Kilovolt
Kw	Kilowatt
Kwh	Kilowatt Hour
LAER	Lowest Achievable Emission Rate
LAI	Leaf Area Index
LANDSAT	Earth Resources Technology Satellite
LANL	Los Alamos National Laboratory (US)
LAPIO	Low API Oil
LAS	Laser Aerosol Spectrometer
LAS	Linear Alkyl Benzene Sulfonate
LASER	Light Amplification By Stimulated Emission of Radiation
LBL	Lawrence Berkeley Laboratory
LBP	Length Between Perpendiculars
LC	Liquid Chromatography
LC_{50}	Mean Lethal Concentration
LCA	Life-Cycle Assessment
LCC	Life-Cycle Costing
LCI	Life-Cycle Inventory
LCIP	Low Carbon Innovation Programme
LCPD	Large Combustion Plant Directive (EU)
LCR	Lead and Copper Rule
LD_{50}	Mean Lethal Dose
LDA	Laminar Directional Airflow
LDAR	Leak Detection and Repair
LDC	Less Developed Country
LDC	London Dumping Convention
LDF	Local Deposition Fraction
LDH	Lactate Dehydrogenase
LDI	Laser Desorption/Ionization
LDPE	Low Density Polyethylene
LEA	Low Excess Air Operation
LED	Light Emitting Diode
LEL	Lower Explosive Limit
LET	Linear Energy Transfer
LFG	Landfill Gas
LFL	Lower Flammability Limit

LIBS	Laser Induced Breakdown Spectrometry
LID	Low Impact Development
LIDAR	Light Detection and Ranging
LILW-LL	Low and Intermediate Level Waste-Long Lived
LILW-SL	Low and Intermediate Level Waste-Short Lived
LLDPE	Linear Low Density Polyethylene
LLNL	Lawrence Livermore National Laboratory (US)
LLW	Low Level Radioactive Waste
LMC	Lime-Magnesium Carbonate
LMO	Living Genetically Modified Organism
LMx	Low Mix Burner
LNAPL	Light Non-Aqueous Phase Liquid
LNB	Low NOx Burner
LNG	Liquefied Natural Gas
LNT	Linear No-Threshold
LOD	Limit of Deposition
LOEC	Lowest Observable Effects Concentration
LOEL	Lowest Observed Effect Level
LOI	Loss On Ignition
LOOP	Locally Organized and Operated Partnerships
LOPA	Layers of Protection Analysis
LOT	Load On Top
LPC	Limiting Permissible Concentration
Lpcd	Liters Per Capita Per Day
LPG	Liquefied Petroleum Gas
LPS	Lipopolysaccharide
LRT	Long Range Transport Model
LRTAP	Long Range Trans-boundary Air Pollution
LS	Low Sulfur
LSE	London School of Economics
LSHTM	London School of Hygiene and Tropical Medicine
LSI	Langlier Saturation Index
LSM	Land Surface Model
LSTK	Lump-Sum Turnkey
LTAR	Long Term Acceptance Rate
LTVE	Long Tube Vertical Evaporator
LUC	Land Use Classification
LULU	Locally Unacceptable Land Use
LUST	Leaking Underground Storage Tank

LVHV	Low Volume High Velocity
LWL	Low Water Level
LWR	Light Water Reactor
m.t.	Metric Tonne
m.t./yr.	Metric Tonnes/Yr.
M/F	Mass/Food Ratio
$m^3m^{-2}d^{-1}$	Cubic Meters Per Square Meter Per Day
MA_7CD_{10}	Minimum Average 7 Consecutive Day 10 Year Flow
MAA	Mycosporine Amino Acid
MAB	Man and the Biosphere (UNESCO)
MAC	Maximum Acceptable (Allowable) Concentration
MAC(EEC)	Maximum Acceptable Concentration (European Economic Community)
MACS	Miniature Acid-Condensation System
MACT	Maximum Achievable Control Technology
MAFF	Ministry of Agriculture, Fisheries and Food (UK)
MAH	Maleic Anhydride
MANEB	Trimangol 80, Poligram M, Plantineb 80 PM
MANOVA	Multivariate Analysis of Variance
MARAMA	Mid-Atlantic Regional Air Use Management Association (US)
MARC	Major Accident Reporting System (EU)
MARC	Meeting of Acidification Research Coordinators
MARPOL	Convention for the Prevention of Pollution from Ships (IMO)
MAS	Mobile Aerosol Spectrometer
MASER	Microwave Amplification By Stimulated Emission
MATC	Maximum Acceptable Toxicant Concentration
MB	Methylene Blue
MB	Mixed Bed
MBL	Marine Boundary Layer
MBR	Membrane Bioreactor
MBT	Monobutyltin
MBW	Metropolitan Board of Works (UK)
MC	Methyl Chloride
MC	Microcarrier
MCA	Manufacturing Chemists Association
MCA	Multicriteria Analysis
MCDSS	Multicriteria Decision Support System

MCFC	Molten Carbonate Fuel Cell
MCI	Management Charter Initiative (UK)
MCL	Maximum Contaminant Level
MCLG	Maximum Contaminant Level Goal
MCM	Master Chemical Mechanism
MCPA	4,Chloro-2-Methylphenoxyacetic Acid
MCR	National Institute of Materials and Chemical Research (Japan)
MCRT	Mean Cell Residence Time
MCS	Multiple Chemical Sensitivity
MCTT	Multichambered Treatment Train
mdd	$mg \times day^{-1} \times dm^{-2}$
MDG	Millennium Development Goal
MDL	Method Detection Limit
MDS	Multidimensional Scaling
ME	Multiple-Effect Evaporator
MEA	Monoethanolamine
MEA	Multinational Environmental Agreement
MED	Modified Electrodialysis
MED	Multieffect Distillation
MEI	Maximum Exposed Individual
MEK	Methyl Ethyl Ketone
MEMS	Microelectro-Mechanical Systems
MEPA	Meteorology and Environmental Protection Administration (Saudi Arabia)
MEPC	Marine Environment Protection Committee
meq	milliequivalent
meq/L	milliequivalents per Liter
MERV	Minimum Efficiency Reporting Value
MESL	Marine Environmental Studies Laboratory (IAEA)
MEUC	Major Energy Users' Council (UK)
MF	Microfiltration
MFA	Material-Flow Accounting (Analysis)
MFN	Most Favored Nation Status
mg/L	Milligrams Per Liter (ppm) (in water)
MGD	Million Gallons Per Day
MGR	Mobile Genetic Element
MHD	Magnetohydrodynamics
MIBK	Methyl Isobutyl Ketone

MICROMORT	One in a Million Chance of Death From An Environmental Hazard
MIE	Magnetic Ion Exchange
MIE	Minimum Ignition Energy
MIGA	Multilateral Investment Guarantee Agency
MIPAES	Microwave Induced Plasma Atomic Emission Spectroscopy
MIPS	Material Intensity Per Service Unit
MIR	Maximum Incremental Reactions
MIR	Maximum Individual Risk
MIRAGE	CEC Radionuclide Geosphere Migration Project
MJO	Madden-Julian Oscillation
MLD	Mean Lethal Dose
MLD	Million Liters Per Day
MLSS	Mixed Liquor Suspended Solids
MLVSS	Mixed Liquor Volatile Suspended Solids
MM	Methyl Mercaptan
MMA	Methyl Methacrylate
MMC	Monopolies and Mergers Commission (UK)
MMH	Monoethyl Hydrazine
MMR	Maternal Mortality Rate
MMT	Methylcyclopentadienyl Manganese Tricarbonyl
MOA	Memorandum of Agreement
MoA	Ministry of Agriculture (UK)
MoD	Ministry of Defense (UK)
MOEE	Ministry of Environment and Energy (Canada)
MoEF	Ministry of Environment and Forests (India)
MONICA	Monitoring of Trends and Determinants in Cardiovascular Disease
MOU	Memorandum of Understanding (US)
MOX	Mixed Oxide
MPD	Maximum Permissible Dose
MPE	Maximum Permissible Exposure
MPN	Most Probable Number
MPP	Macroporous Polymer
mppcf	Millions of Particles Per Cubic Foot
MPPE	Macroporous Polymer Extraction
MPRSA	Marine Protection, Research and Sanctuaries Act (Ocean Dumping Act) (US)

MR	Magnetic Resonance
MR	Mass Removal
MRC	Medical Research Council (UK)
MRDL	Maximum Residual Disinfectant Level
MRDLG	Maximum Residual Disinfectant Level Goal
MRF	Material Recovery Facilities
MRF	Municipal Recycling Facilities
MRI	Mean Recurrence Interval
MRS	Marine Remote Sensing
MRS	Monitored Retrievable Storage
MS	Management System
MS	Mass Spectrometry
MS	Mobile Source
MSAT	Mobile Source Air Toxics
MSDS	Material Safety Data Sheets (US)
MSE	Mean Square Estimate
MSF	Multistage-Flash Evaporator
MSG	Monosodium Glutamate
MSHA	Mine Safety and Health Administration (US)
MSL	Mean Sea Level
MSMR	Mean Standardized Mortality Rate
MSS	Multispectral Scanner
MSW	Municipal Solid Waste
MSY	Maximum Sustainable Yield
Mta^{-1}	Million Tons Per Year (Annum)
MTB	Multiple Twinned Particles
MTBE	Methyl Tert(iary)-Butyl Ether
MTBF	Mean Time Between Failures
MtCeq	Million Tons Carbon Equivalent
MTD	Maximum Tolerable Dose
MTG	Methanol to Gasoline
MTHF	Methyltetrahydrofuran
MTI	Mixture Toxicity Index
mtoe	million tonnes oil equivalent
mtpa	million tonnes per annum
MUA	Municipal Utilities Authority
MVC	Mechanical Vapor Compression
MW	Megawatt
MW	Monitoring Well

MWCO	Molecular Weight Cutoff
MWT	Multiple Well Tracer Test
N	Newton
NA	Nutrient Agar
NAA	Neutron Activation Analysis
NAAQS	National Ambient Air Quality Standards (US)
NACA	National Agricultural Chemicals Association (US)
NAD	Nitric Acid Dihydrate
NADH	Nicotinamide Adenine Dinucleotide
NADP/NTN	National Atmospheric Deposition Program/National Trends Network
NADW	North Atlantic Deep Water
NAE	National Academy of Engineering (US)
NAFO	North Atlantic Fisheries Organization
NAFTA	North American Free Trade Agreement
Nagra	Waste Disposal Organization (Switzerland)
Naics	North American Industry Classification System
NaNp	Sodium Napthalenide
NAO	National Audit Office (UK)
NAO	North Atlantic Oscillation
NAP	National Allocation Plan (UK) (EU)
NAPAG	National Academics Policy Advisory Group (UK)
NAPAP	National Acid(ic) Precipitation Assessment Program (US)
NAPCA	National Air Pollution Control Administration (US)
NaPEG	Sodium Polyethylene Glycol
NAPL	Non-Aqueous Phase Liquid
NAS	National Academy of Sciences (US)
NAS	National Audubon Society
NASA	National Aeronautics and Space Administration (US)
NAST	National Assessment Synthesis Team
NAST	National Assessment Synthesis Team (US)
NAT	Nitric Acid Trihydrate
NB	Nitrobenzene
NB	Nutrient Broth
NBOD	Nitrogenous Biochemical Oxygen Demand
NBS	National Bureau of Standards (US)
NC	Number Concentration
NCAB	National Cancer Advisory Board (US)
NCAR	National Center for Atmospheric Research (US)

NCB	National Coal Board (UK)
NCC	Nature Conservancy Council (UK)
NCDC	National Climate Data Center
NCEA	National Center For Environmental Assessment (US)
NCEES	National Council of Examiners for Engineers and Surveyors (US)
NCEP	National Centers for Environmental Prediction (US)
NCHGR	National Center for Human Genome Research (US)
NCHS	National Center for Health Statistics (US)
NCI	National Cancer Institute (US)
NCI-MS	Negative Chemical Ionization–Mass Spectrometry
NCP	National Contingency Plan (US)
NCRP	National Council on Radiation Protection and Measurements (US)
NCS	Notification of Compliance Status (US)
ND	Not (None) Detected (Detectable)
NDA	New Drug Approval (Application)
NDA	Nuclear Decommissioning Authority (UK)
NDIR	Nondispersive Infrared
NEA	Nuclear Energy Agency (UK)
NEAT	National Environmental Achievement Award
NECI	Network Coordinating Institute
NEDO	New Energy and Industrial Technology Development Organization (Japan)
NEERI	National Environmental Engineering Research Institute (India)
NEF	Noise Exposure Forecast
NEL	No Effect Level
NEP	National Energy Policy (US)
NEPA	National Environmental Policy Act (US)
NEPA	National Environmental Protection Agency (PRC)
NERC	National Environmental Research Council (UK)
NERI	National Environmental Research Institute (Denmark)
NERI	Nuclear Energy Research Initiative
NERL	National Exposure Research Laboratory (US)
NESCAUM	Northeast States for Coordinated Air Use Management (US)
NESDIS	National Environmental Satellite Data and Information Center

NESHAP	National Emissions Standards for Hazardous Air Pollutants (US)
NESS	National Environmental Satellite Service (US)
NETA	New Electricity Trading Agreements (UK)
NETCEN	National Environmental Technology Centre (UK)
NETL	National Energy Technology Laboratory (US)
NEXRAD	Next Generation Weather Radar
NF	Nanofiltration
NFFO	Non-Fossil Fuel Obligation
NFIP	National Flood Insurance Program (US)
NFMA	National Forest Management Act (US)
NFPA	National Fire Protection Association (US)
NFSOM	Near-Field Scanning Optical Microscopy
NSPE	National Society of Professional Engineers (US)
NGL	Natural Gas Liquid
NGO	Non-Government Organization (UN)
NGV	No Guideline Value
NGV	Natural Gas Vehicle
NHC	National Hurricane Center (US)
NHEERL	National Health and Environmental Effects Research Laboratory (US)
NHER	National Home Energy Rating (UK)
NHLBI	National Heart, Lung and Blood Institute (US)
NHS	National Health Service (UK)
NI	Near-Infrared
NIAID	National Institute of Allergy and Infectious Disease (US)
NIBS	National Institute of Building Sciences (US)
NICOLE	Network for Industrially Contaminated Land in Europe
NIEHS	National Institute of Environmental Health Sciences (US)
NIH	National Institute of Health (US)
NII	Nuclear Installations Inspectorate (UK)
NIMBY	"Not In My Backyard"
NIOSH	National Institute for Occupational Safety and Health (US)
NIREX	Nuclear Industry Radioactive Waste Executive (UK)
NIST	National Institute of Standards and Technology (US)
NL/h	Normal Liters per Hour
NLER	Nitrogen-Limited Enrichment

NLM	National Library of Medicine
NMC	National Meteorological Centre (UK)
NMD	Number Mean Diameter
NMFS	National Marine Fisheries Service (US)
NMHC	Non-Methane Hydrocarbon
NMMAPS	National Morbidity, Mortality and Air Pollution Study (US)
NMOC	Non-Methane Organic Carbon
NMR	Neonatal Mortality Rate
NMR	Nuclear Magnetic Resonance
NMVOC	Non-Methane Volatile Organic Compound
NNI	Noise and Number Index
NNR	National Nature Reserve (UK)
$NO_{2,3}$	Nitrite Plus Nitrate
NOAA	National Oceanographic and Atmospheric Administration (US)
NOAEL	No Observable Adverse Effect Level
NOEC	No Observable Effects Concentration
NOEL	No Observed Effect Level
NOHSCP	National Oil and Hazardous Substance Contingency Plan (US)
NOM	Natural Organic Matter
NONHEL	Non Highly Erodible Land
NOS	National Occupational Standard
NOS	National Ocean Survey (US)
NO_x	Nitrogen Oxides
NO_y	Reactive Nitrogen
NP	Neutralization Point
NPAA	Noise Pollution and Abatement Act (US)
NPCA	National Parks and Conservation Association (US)
NPDES	National Pollutant Discharge Elimination System (US)
NPL	National Priority List (US)
NPP	Net Primary Production (Productivity)
NPP	Nonprecipitated Phosphorous
NPP	Nuclear Power Plant
NPR	National Performance Review (US)
NPRI	National Pollution Release Inventory (Canada)
NPRM	Notice of Proposed Rule Making (US)
NPS	National Park Service (US)

NPSH	Net Positive Suction Head
NRA	National Rivers Authority (UK)
NRC	Nuclear Regulatory Commission (US)
NRC	National Research Council (Canada) (US)
NRCS	Natural Resources Conservation Service
NRD	Natural Resources Damage
NRDA	Natural Resource Damage Assessment
NRDC	Natural Resources Defense Council
NREL	National Renewable Energy Laboratory (US)
NRMRL	National Risk Management Research Laboratory (US)
NRPB	National Radiological Protection Board (UK)
NRR	Net Reproduction Rate
NRT	National Response Team (US)
NSERC	National Science and Engineering Research Council (Canada)
NSF	National Science Foundation (US)
NSPS	New Source Performance Standards (US)
NSR	New Source Review
NSTS	North Sea Task Force
NTA	Nitrilotriacetic Acid
NTI	National Toxics Inventory (US)
NTI	Nuclear Threat Initiative
NTN	National Trends Network
NTP	National Toxicology Program (US)
NTS	Not To Scale
NTSB	National Transportation Safety Board (US)
NTTC	National Technology Transfer Center (US)
NTU	Nephelometric Turbidity Unit
NTU	Number of Transfer Units
Nu	Nusselt Number
NURP	National Urban Runoff Program (US)
NWP	Numerical Weather Prediction
NWR	National Wildlife Refuge (US)
NWS	National Weather Service (US)
NWTRB	Nuclear Waste Technical Review Board (US)
O&M	Operation and Maintenance
O.D.	Optical Density
O8CDD	Octochloro Dibenzo-p-Dioxin
O8CDF	Dibenzofuran

OA	Overfire Air
OBO	Ore-Bulk-Oil
OC	Organic Carbon
OCAG	Off-Site Consequences Analysis Guidance (US)
OCCM	Office of Air Quality Planning and Standards Control Cost Manual (US)
OCGCM	Ocean Coupled General Circulation Model
OCP	Organopesticides
OCPSF	Organic Chemicals, Plastics and Synthetic Fibers
OCS	Outer Continental Shelf
ODA	Overseas Development Agency (UK)
ODP	Ocean Drilling Programme (Australia)
ODP	Ozone Depletion Potential
ODS	Ozone Depleting Substance
OEAS	Oxygen Enriched Air Staging
OECD	Organization for Economic Cooperation and Development
OECF	Overseas Economic Cooperative Fund (Japan)
OED	Operations Evaluation Department (World Bank)
OEL	Occupational Exposure Limit
OEM	Office of Emergency Management (US)
OEM	Original Equipment Manufacturer
OERR	Office of Emergency and Remedial Response (US)
OES	Occupational Exposure Standard (UK)
OES	Office of Endangered Species (US)
OFA	Overfire Air
OH&S	Occupational Health and Safety
OHPA	Obligatory Hydrogen-Producing Acetogenic
OHS	Occupational Hygiene Secretariat (UK)
OIE	Organisation Internationale des Epizooties
OIES	Oxford Institute for Energy Studies (UK)
OLED	Organic Light-Emitting Diode
OLR	Organic Loading Rate
OMB	Office of Management and Budget (US)
OMG	Old Mixed Grade
OMT	Open Market Trading
OMTR	Open Market Trading Credits Rule (US)
OMZ	Oxygen-Minimum Zone
ONBC	Overnight Nutrient Broth Culture
ONDRAF	Waste Disposal Organization (Belgium)

OP	Organophosphate
OPA	Oil Pollution Act (US)
OPCW	Organization for Prevention of Chemical Warfare
OPEC	Organization of Petroleum Exporting Countries
OPETS	Organization for the Promotion of Energy Technologies (EU)
OP-FTIR	Open-Path Fourier Transform Infrared
OPLA	High Level Waste Disposal Program (Holland)
OPS	Office of Pipeline Safety (US)
ORD	Office of Research and Development (US)
ORI	Office of Research Integrity (US)
ORNL	Oak Ridge National Laboratory (US)
ORP	Oxidation-Reduction Potential
ORSANCO	Ohio River Sanitation Commission (US)
OSAT	On-Site Assistance Team (EC)
OSHA	Occupational Safety and Health Act (US)
OSPAR	Oslo-Paris Agreement
OSPE	Ontario Society of Professional Engineers
OSPREY	Ocean Swell Powered Renewable Energy (UK)
OST	Office of Science and Technology (UK) (US)
OSW	Office of Saline Water (US)
OSWER	Office of Solid Waste and Emergency Response (US)
OTA	Office of Technology Assessment (US)
OTEC	Ocean Thermal Energy Conversion
OTP	Ozone Trends Panel
OTR	Oxygen Transfer Rate
OTR	Ozone Transport Region (US)
OUR	Oxygen Uptake Rate
OVA	Organic Vapor Analysis
OW	Oil in Water
Ox	Total Oxidants
OY	Optimum Yield
P	Polystyrene
P and T	Pump and Treat
P ENG	Professional Engineer (Canada)
p.a.	Per Annum
P^2	Pollution Prevention
Pa	Pascal
PA	Polyamide

PABA	Para-Amino-Benzoic Acid
PAC	Polyaluminum Chloride
PAC	Powdered Activated Carbon
PACl	Polyaluminum Chloride
PACT	Programme for Alternative Fluorocarbon Toxicity Testing
PAFC	Phosphoric Acid Fuel Cell
PAH	Polynuclear Aromatic Hydrocarbons
PAHO	Pan American Health Organization (WHO)
PAL	Plantwide Applicability Limits
PAN	Dithiozone
PAN	Peroxyacetyl Nitrate
PAN	Pesticide Action Network
PARC	Pan African Rinderpest Campaign
PARCCS	Precision, Accuracy, Representativeness, Comparability, Completeness, Sensitivity
PARCOM	Paris Commission
PARIS	Program for Assisting the Replacement of Industrial Solvents
PATH	Plan for Analyzing and Testing Hypotheses
PBDD	Polybrominated Dioxins
PBL	Planetary Boundary Layer
PBMR	Pebble-Bed Modular Reactor
PBS	Package Boiler Simulator
PC	Polycarbonate
PC	Pulverized Coal
PCA	Plate Count Agar
PCA	Principal Component Analysis
PCB	Polychlorinated Biphenyl
PCBP	Polychloro Bi Phenylene
PCBz	Polychloro Benzene
PCCY	Polychlorinated Chrysene
PCD	Particle Charge Detector
PCD	Process Control Diagram
PCDD	Polychlorodibenzo-p-Dioxin
PCDET	Post-Completion Discrete Extraction Test (US)
PCDF	Polychlorinated Dibenzofuran
PCDPE	Polychlorinated Diphenyl Ether
PCE	Perchlorethylene
PCE	Tetrachloroethylene

pCi/L	Picocuries Per Liter
PCN	Polychlorinated Napthalene
PCP	Pentachlorophenol
PCPA	Post-Closure Performance Assessment (US)
PCPY	Polychlorinated Pyrene
PCQ	Polychlorinated Quaterphenyl
PCQE	Polychlorinated Quaterphenyl Ether
PCR	Polymerase Chain Reaction
PCT	Polychlorinated Triphenol
PCV	Positive Crankcase Ventilation
PDA	Potato Dextrose Agar
PDE	Partial Differential Equation
PDF	Probability Density Function
PDO	Pacific Decadal Oscillation
PE	Performance Evaluation
PE	Polyethylene
PE	Professional Engineer (US)
PEA	Performance Evaluation Audit
PEACE	Pollution Effects on Asthmatic Children in Europe
PEC	Process Economic Program
PEELS	Parallel Acquisition Systems
PEF	Pulsed electric field
PEFR	Peak Expiratory Flow Rate
PEI	Potential Environmental Impact
PEIs	Professional Engineering Institutions (UK)
PEL	Permissible Exposure Limit
PEM	Proton Exchange Membrane
PEMFC	Proton Exchange Membrane Fuel Cell
PEN	Polyethylene Napthalate
PEO	Professional Engineers Ontario
PER	Partial Exfiltration System
PES	Project Environmental Summary
PES	Public Electricity Supplier
PET	Polyethylene Terephthalate
PF	Phenol-Formaldehyde
PF	Pulverized FuelP5CB 3,3′,4,4′,5-Pentachlorinated Biphenyl
PFBC	Pressurized Fluidized Bed Combustion
PFC	Perfluoro (Carbon) Compounds

PFD	Process Flow Diagram
PFOA	Perfluorooctanoic Acid
PFR	Plug Flow Reactor
PFRP	Process To Further Reduce Pathogens
PFS	Pulverized Fly Ash
PFT	Peak Flame Temperature
PHA	Process Hazard Analysis
PHA	Pulse Height Analysis
PHARE	CEC Program to Assist Eastern European Countries
PHB	Polyhydroxy Butyrate
PHC	Petroleum Hydrocarbons
PHE	Public Health Engineer (UK)
PHS	US Public Health Service
PI	Principal Investigator
PIC	Prior Informed Consent Procedure
PIC	Product of Incomplete Combustion
PIEL	Pharmacologically Insignificant Exposure Limit
PIRG	Public Interest Research Group (US)
PIXIE	Particle Induced X-Ray Emission
PL	Public Law (US)
PLA	Polylactic Acid
PLC	Polylimonene Carbonate
PLC	Programmable Logic Controller
PLL	Probable Loss of Life
PLM	Polarized Light Microscopy
PM	Particulate Matter
PM0.1	Ultrafine Particle
PM10	Particles Less Than 10 Micrometers
PM2.5	Particles Less Than 2.5 Micrometers
PMA	Phenyl Mercuric Acetate
PMM	Polymethyl Methacrylate
PMMA	Polymethylmethacrylate
PMN	Phenyl Mercuric Nitrate
PMN	Polymorphonuclear Neutrophils
PMN	Premanufacture Notice (US)
PNC	Japanese Nuclear Power Corporation
PNdB	Perceived Noise Level
PNGV	Partnership for a New Generation of Vehicles
PNL	Pacific Northwest Laboratory (US)

PNNL	Pacific Northwest National Laboratory (US)
PODAAC	Physical Oceanography Distributed Active Archive
POE	Point of Entry
POHC	Principal Organic Hazardous Constituent
POL	Project Objectives Letter (US)
PolyTHF	Polytetrahydrofuran
POM	Polycyclic Organic Matter
POM	Princeton Ocean Model
POP	Persistent Organic Pollutant
POT	Peaks-Over Threshold
POTW	Publicly Owned Treatment Works (US)
POU	Point of Use
PP	Pilot Plant
PP	Polypropylene
PP	Precautionary Principle
ppb	Parts Per Billion
ppbv	Parts Per Billion by Volume
PPE	Personal Protective Equipment
PPEQ	Pre-Project Environmental Questionnaire
ppm	Parts Per Million (Milligrams Per Liter) (Grams Per Cubic Meter)
ppmv	Parts Per Million by Volume
ppmw	Parts Per Million by Weight
PPP	Polluter Pays Principle
ppt	Parts Per Trillion
ppth	Parts Per Thousand
PPU	Pertinent Process Unit
Pr	Prandtl Number
PRA	Paperwork Reduction Act (US)
PRA	Probabilistic Risk Assessment
PRAS	Prereduced Anaerobically Sterilized
PRG	Preliminary Remediation Goals (US)
PRP	Potentially Responsible Party (US)
PRZ	Potential Repository Zone
PS	Polystyrene
PSA	Probabilistic Safety Assessment
PSAC	President's Science Advisory Committee (US)
PSC	Polar Stratospheric Cloud
PSD	Prevention of Significant Deterioration (US)

PSF	Peat-Sand Filter
PSF	Pounds Per Square Foot
PSI	Pollutant Standard Index
PSI	Pounds Per Square Inch
PSM	Process Safety Management (US)
PSR	Preliminary Safety Report (Review)
PSRP	Process To Significantly Reduce Pathogens
PTF	Permanent Threshold Shift
PTFE	Polytetrafluoroethylene
PTM	Photochemical Trajectory Model
PTR	Project Tracking Register
PTWI	Provisional Tolerable Weekly Intake
PU	Polyurethane
PULSAR	Phillips' Ultra Low Sulfur Atmospheric Residue
PV	Parametric value
PV	Permanganate Value
PV	Photovoltaic
PV	Polyvinyl
PVA	Polyvinyl Acetate
PVA	Process Vulnerability Analysis
PVC	Polyvinyl Chloride
PVS	Physical Vapor Synthesis
PVT	Photovoltaic/Thermal
PWS	Public Water System
PZC	Point of Zero Charge
QA	Quality Assurance
QAC	Quaternary Ammonium Compounds
QAPP	Quality Assurance Project Plan (US)
QC	Quality Control
QEP	Qualified Environmental Professional
QF	Qualified Facility
QIP	Quality Improvement Program
QP	Qualified Person
QPF	Quantitative Precipitation Forecast
QRA	Quantitative Risk Assessment
QSAR	Quantitative Structure Activity Relationship
QUANGO	Quasi Autonomous Non-Governmental Organization
QUARG	Quality of Urban Air Review Group
R	Rankine

R	Universal Gas Constant
r	Roentgen
R&D	Research and Development
R/P	Reserve/Production
RA	Rapid Appraisal
RAB	Registrar Accreditation Board
RACT	Reasonably Available Control Technology
Rad	Roentgen-absorption-dose
RADWASS	IAEA Publication on Radioactive Waste
RAF	Radiation Amplification Factor
RAMP	Regional Air Management Partnership
RAMS	Regional Atmospheric Model System
RAP	Rapid Assessment Program
RAR	Reasonable Assumed Resource
RaSoS	Raman Sort Spectrometer
RBC	Rotating Biological Contactor
RBCA	Risk Based Corrective Action
RBE	Relative Biological Effectiveness
RBSL	Risk-Based Screening Levels
RCEP	Royal Commission on Environmental Pollution (UK)
RCM	Regional Climate Modelling
RCO	Regenerative Catalytic Oxidation
RCRA	Resource Conservation and Recovery Act (US)
RCT	Randomized Controlled Trial
RCT	Reference Control Technology
RD&D	Research, Development and Demonstration
RDF	Refuse-Derived Fuel
RDS	Residue-Oil Hydrodesulfurization
Re	Reynolds Number
REACH	Registration, Evaluation, Authorization of Chemicals
RECLAIM	Regional Clean Air Incentives Market (US)
REEEP	Renewable Energy and Energy Efficiency Partnership
REEF	Radiation Effects Research Foundation
REI	Regional Environmental Initiative
REIO	Regional Economic Integration Organization
REL	Recommended Exposure Limit
Rem	Roentgen-equivalent-man
REMA	Regulatory Environmental Modelling of Antifoulants
REO	Renewable Energy Obligation

Rep	Roentgen-equivalent-physical
RF	Radio Frequency
RFBR	Russian Federation for Basic Research
RFCC	Residue-Oil Fluid Catalytic Cracking
RFF	Resources For The Future
RFG	Reformulated Gasoline
RFLPS	Restriction-Fragment-Length Polmorphisms
RGH	Renewably Generated Hydrogen
RGR	Relative Growth Rate
RH	Relative Humidity
RHA	Regional Health Authority (UK)
RI	Remedial Investigation
RI	Return Interval
Ri	Richardson Number
RIA	Regulatory Impact Analysis
RIBA	Royal Institution of British Architects
RI-FS	Remedial Investigation and Feasibility Study (US)
RIIA	Royal Institute of International Affairs (UK)
RL50	Residue Half Life
RMCL	Recommended Maximum Contaminant Level
RME	Reasonable Maximum Exposure
RMLT	Regression Models in Life Tables
RMP	Risk Management Program (Plan) (US)
RMR	Required Mass Removal
RNA	Ribonucleic Acid
RNI	Rate of Natural Increase
RO	Renewables Obligation
RO	Reverse Osmosis
ROC	Renewable Obligation Certificate
ROD	Record of Decisions (Superfund) (US)
ROFA 6	Residual Oil Fly Ash (No. 6 Fuel Oil)
RORO	Roll On Roll Off
RoSPA	Royal Society for the Prevention of Accidents (UK)
RP	Return Period
RPA	Radiation Protection Act (UK)
RPA	Renewable Power Association
RPE	Respiratory Protection Equipment
RPP	Radiation Protection Program (Canada)
RPS	Renewable Portfolio Standard

RQ	Reportable Quantity (US)
RRRR	Reduce/Recovery/Recycle/Reuse
RRT	Relative Retention Time
RS	Royal Society (UK)
RSA	Radioactive Substances Act (UK)
RSC	Reactor Safety Commission (Germany)
RSC	Records of Site Condition
RSC	Royal Society of Chemistry (UK)
RSD	Relative Standard Deviation
RSH	Royal Society of Health (UK)
RTD	Residence Time Distribution
RTDF	Remediation Technologies Development Forum
RTO	Regenerative Thermal Oxidation (Oxidizer)
RTP	Research Triangle Park
RVP	Reid Vapor Pressure
RWB	Regional Water Board
RWMAC	Radioactive Waste Management Advisory Committee (UK)
RWQCB	Regional Water Quality Control Board
S and T	Science and Technology
S&DSI	Stiff and Davis Stability Index
S/S	Solidification/Stabilization
SAED	Selected Area Electron Diffraction
SAGE	Stratospheric Aerosol and Gas Experiment
SALR	Saturated Adiabatic Lapse Rate
SAR	Sodium Adsorption Ratio
SAR	Specific Absorption Rate
SAR	Structure-Activity Relationship
SARA	Superfund Amendments and Reauthorization Act (US)
SARTOR	Standards and Routes to Registration (UK)
SASS	Source Assessment Sampling System
SAVEII	Specific Action for Vigorous Energy Efficiency (EU)
SBC	Submerged Biological Contactor
SBEM	Simplified Building Energy Model
SBR	Sequencing Batch Reactor
SBS	Sick Building Syndrome
SBT	Segregated Ballast Tank
Sc	Schmidt Number
SC	Sierra Club

SCADA	Supervisory Control and Data Acquisition
SCAPS	Site Characterization and Analysis Penetrometer System
SCAQMD	South Coast Air Quality Management District
SCF	Standard Cubic Feet
SCF	Supercritical Fluid
SCOP	Subcommittee On Oil Pollution
SCOPE	Scientific Committee on Problems of the Environment
SCPH	Sodium Carbonate Peroxyhydrate
SCR	Selective Catalytic Reduction
SCS	Soil Conservation Service (US)
SCW	Supercritical Water
SCWO	Supercritical Wet (Water) Oxidation
SD	Solar Distillation
SD	Standard Deviation
SD	Sustainable Development
SDBS	Sodium Dodecyl Benzene Sulfonate
SDI	Silt Density Index
SDR	Surplus Discrete Reductions (US)
SDWA	Safe Drinking Water Act (US)
SEA	Single European Act (EU)
SEA	Strategic Environmental Assessment
SEARR	South East Asia Rainforest Research Program
SEASAT	Earth Satellite for Sea Surveys
SEC	Site-Specific Energy Consumption
SEC	Size Exclusion Chromatography
SED	Soil Evacuation and Disposal Plan (US)
SEE	Society of Environmental Engineers (UK)
SEER	Surveillance, Epidemiology and End Result
SEFI	European Society for Engineering Education
SELCHP	South East London Combined Heat and Power (UK)
SEM	Scanning Electron Microscope
SEP	Supplemental Environmental Project
SERC	Scientific and Engineering Research Council (UK)
SERI	Solar Energy Research Institute
SET	Science, Engineering, Technology
SETAC	Society of Environmental Toxicology and Chemistry
SF	Safety Factor
SF	Sub-surface flow
SFIP	Sector Facility Indexing Project (US)

SG	Sustained Growth
Sh	Sherwood Number
SHASE	Society of Heating, Air Conditioning and Sanitary Engineers (Japan)
SHE	Safety, Health and the Environment
SHE	Systeme Hydrologique Europeen
SHP	Shaft Horsepower
SHPO	State Historic Preservation Officer (US)
SHWP	Seismic Hazards Working Party
SI	International System of Units (Systeme International d'Unites)
SI	Saturation Index
SI	Statutory Instruments (UK)
SIC	Standard Industrial Classification (US)
SIL	Safety Integrity Level
SIM	Selective Ion Monitoring
SIMS	Secondary Ion Mass Spectrometry
SIP	Site Implementation Plan (US)
SIP	Sterilization In Place
SIPI	Scientists' Institute for Public Information
SIS	Susceptible-Infected-Susceptible
SKAPP	Scientific Knowledge and Public Policy
SKI	Radiation Protection Institute (Sweden)
SL	Surface Layer
SLAMM	Source Loading and Management Model
SLAPP	Strategic Lawsuit Against Public Participation (Canada)
SLM	Supported Liquid Membrane
SMB	Simulated Moving Bed
SMCL	Secondary Maximum Contaminant Level
SMCRA	Surface Mining Control and Reclamation Act (US)
SME	Solar Mesospheric Explorer
SMOW	Standard Mean Ocean Water
SMP	Soluble Microbial Product
SMPS	Scanning Mobility Particle Sizer
SMR	Standardized Mortality Ratio (Rate)
SMZ	Surface Modified Zeolite
SNAP	Significant New Alternatives Policy
SNARL	EPA Suggested No Adverse Response Level (US)
SNCR	Selective Noncatalytic Reduction

SNL	Sandia National Laboratory (US)
SNL	Scottish Nuclear Ltd
SNL	Sense Noise Level
SOA	Secondary Organic Aerosols
SOC	Soluble Organic Compound (Chemical)
SOC	Synthetic Organic Chemical
SOCMI	Synthetic Organic Chemical Manufacturing Industry
SOD	Sediment Oxygen Demand
SOFC	Solid Oxide Fuel Cell
SOI	Southern Oscillation Index
SoPHE	Society of Public Health Engineers
SOx	Sulfur Oxides
SPCC	Spill Prevention Control and Countermeasures
SPE	Solid Phase Extraction
SPE	Survivor Population Effect
SPFC	Solid Polymer Fuel Cell
SPI	Society of the Plastics Industry
SPM	Scanning Probe Microscopy
SPME	Solid-Phase Microextraction
SPOT	Systeme Pour l'Observation de La Terre (France)
SPSS	Sulfur Polymer Stabilization and Solidification
SR&O	Statutory Regulations and Orders (UK)
SRC	Science Research Council (UK)
SREL	Savannah River Ecology Laboratory
SRF	State Revolving Fund (US)
SRT	Solids Retention Time
SS	Suspended Solids
SSA	Specific Surface Area
SSBY	Sewage Solids By-Product
SSD	Safe Separation Distance
SSHRC	Social Sciences and Humanities Research Council (Canada)
SSI	Nuclear Safety Inspectorate (Sweden)
SSIT	Society On Social Implications of Technology
SSOW	Source Separated Organic Waste
SSSI	Site of Special Scientific Interest (UK)
SST	Sea-Surface Temperature
SST	Supersonic Transport
SSTL	Site Specific Target Levels

STAPPA	State and Territorial Air Pollution Program Administrators (US)
STEL	Short Term Exposure Limit
STEM	Standard Transmission Electron Microscope
STEM	Science, Technology, Engineering, Mathematics
STEP	Septic Tank Effluent Pump
STEP	Standard for Exchange of Production
STM	Scanning Tunneling Microscopy
STORM	Storage Treatment Overflow and Runoff Model
STP	Standard Temperature and Pressure
STS	Supercooled Ternary Solution
SU	Standard Unit
SUVA	Specific Ultraviolet Absorbance
Sv	Sievert
SVE	Soil Vapor Extraction
SVI	Sludge Volume Index
SVIN	Swiss Association of Women Engineers
SVOC	Semivolatile Organic Compound
SVR	Sludge Volume Ratio
SWAMP	Storm Water Assessment, Monitoring and Performance Program (Canada)
SWDA	Solid Waste Disposal Act (US)
SWE	Society of Women Engineers (US)
SWERA	Solar and Wind Energy Resource Assessment
SWFGD	Seawater Flue Gas Desulfurization
SWH	Solar Water Heater
SWJ	Sewage Works Journal
SWMM	Stormwater Management Model
T4CB	3,3′,4,4′-Tetrachlorinated Biphenyl
TA	Total Alkalinity
TAC	Total Annual Cost
TADC	Tire-Derived Activated Carbon
TAEE	Tert-Amyl Ethyl Ether
TAME	Tert-Amyl Methyl Ether
TAO	Tropical Atmosphere Ocean
TAPPI	Technical Association of the Pulp and Paper Industry
TAR	Tribal Authority Rule
TAW	Technical Advisory Committee on Water Defenses (Holland)

TB	Tracheobronchial
TBT	Tributyltin
TC	Total Coliforms
TCA	Trichloroethane
TCDD	2,3,7,8-Tetrachlorodibenzo-p-Dioxin
TCDF	Tetrachlorodibenzofuran
TCE	Trichloroethylene
TCER	Target Carbon Emission Rate
TCF	Totally Chlorine Free
TCI	Total Capital Investment
TCLP	Toxicity Characteristic Leaching Procedure (Potential) (US)
TCN	Technical Cooperation Network
TCP	Trichloropropane
TCP/IP	Transmission Control Protocol/ Internet Protocol
TCR	Total Coliform Rule
TCU	True Color Unit
TDH	Total Dynamic Head
TDI	Tolerable Daily Intake
TDI	Toluene-2,4-Diisocyanate
TDS	Total Dissolved Solids
TEA	Triethanolamine
TEB	Total Exchangeable Bases
TECSEC	Technical Secretariat
TEL	Tetraethyl Lead
TEL	Total Energy Line
TEM	Total Extractable Matter
TEOM	Tapered Elemental Oscillating Microbalance
TEQ	Toxic Equivalent
TEWI	Total Equivalent Warming Impact
TFAP	Tropical Forestry Action Plan
TFE	Tetrafluoroethylene
THAA	Total Haloacetic Acids
THC	Thermohaline Circulation
THC	Total Hydrocarbon
THESUS	Thermal Solar European Power Station
THF	Tetrahydrofuran
THM	Chloro-Organics
THM	Thermo-Hydro-Mechanical

THM	Trihalomethane
ThOD	Theoretical Oxygen Demand
THORP	Thermal Oxide Reprocessing Plant
TI	Tobacco Institute
TIC	Total Industry Control
TIC	Total Inorganic Carbon
TIE	Toxicity Investigation Evaluation
TIEL	Toxicologically Insignificant Exposure Limit
TIMES	Thermoelectric Integrated Membrane Subsystem
TIO	Technology Innovation Office (US)
TISE	"Take It Somewhere Else"
TJ	Tetrajoule
TKN	Total Kjeldahl Nitrogen
TL	Threshold Limit
TL,m	Threshold (Tolerance) Limit, median
TLC	Total Lung Capacity
TLV	Threshold Limit Value
TM	Thematic Mapper
TMC	Total Monomer Concentration
TMDL	Total Maximum Daily Load
TMI	Three Mile Island
TMP	Transmembrane Pressure
TN	Total Nitrogen
TNCB	Trinitrochlorobenzene
TNO	Netherlands Organization for Applied Scientific Research
TNO	Total Number of Organisms
TNRCC	Texas Natural Resources Conservation Commission
TOA	Tropical Ocean Atmosphere
TOC	Total Organic Carbon
toe	tonnes oil equivalent
TOGA	Tropical Ocean Global Atmosphere
TOMS	Total Ozone Mapping Spectrometer
TOVALOP	Tank Owners Voluntary Agreement On Liability For Oil Pollution
TOX	Tetradichloroxylene
TP	Total Phosphorous
TPA	Tetraphthalic Acid
TPH	Total Petroleum Hydrocarbons

TPL	Trust for Public Lands
TPP	Tripolyphosphate
TPU	Thermal Processing Unit
TPV	Third-Party Verification
TQ	Threshold Quantity
TQM	Total Quality Management
TRE	Total Resource Effectiveness
TRE	Toxicity Reduction Evaluation
TRI	Toxics Release Inventory (US)
TRIP	Trade Related Intellectual Property
TRS	Total Reduced Sulfur
TS	Total Solids
TSA	Technical Systems Audit
TSCA	Toxic Substances Control Act (US)
TSCATS	Toxic Substances Control Act Test Submission
TSDF	Treatment, Storage and Disposal Facility
TSDHW	Transportation, Storage and Disposal of Hazardous Wastes
TSE	Transmissible Spongiform Encephalpathies
TSP	Total Soluble Phosphorous
TSP	Total Suspended Particulates
TSS	Total Suspended Solids
TT	Treatment Technique
TTHM	Total Trihalomethane
TTN	Technology Transfer Network (US)
TTNBS	Technology Transfer Network Bulletin System
TTO	Total Toxic Organics
TTO	Troposphere Ozone
TTS	Temporary Threshold Shift
TTT&O	Time, Temperature, Turbulence and Oxygen
TUC	Trades Union Congress (UK)
TVC	Thermal Vapor Compression
TVO/IVO	Finnish Nuclear Organization
TWA	Thames Water Authority
TWA	Time Weighted Average
TWh	Tetrawatt Hours
UAM	Urban Airshed Model
UAMV	Urban Airshed Model V
UASB	Upflow Anaerobic Sludge Blanket (Bed)
UATI	Union of Technical Associations and Organizations

UATS	Urban Air Toxics Strategy
UATW	Unified Air Toxics Web Site
UBA	German Federal Environmental Agency
UC	Uniformity Coefficient
UCM	Unresolved Complex Mixtures
UCPC	Ultrafine Condensation Particle Counter
UCS	Unconditioned Stimulus
UEIP	Use and Exposure Information Voluntary Project (US)
UEL	Upper Explosive Limit
UF	Ultrafiltration
UFL	Upper Flammability Limit
UKMO	UK Meteorological Office
UKPHA	UK Public Health Association
UL	Underwriters Laboratories
ULCC	Ultra Large Crude Carrier
ULFT	Ultra Low-Flow Toilet
ULSD	Ultra Low Sulfur Diesel
ULSP	Ultra Low Sulfur Petrol
UMIST	University of Manchester Institute of Science and Technology
UMRA	Unfunded Mandates Reform Act (US)
UN	United Nations
UNCED	UN Conference on Environment and Development
UNCED	UN Convention on Environmental Diversity
UNCHE	UN Conference On The Human Environment
UNCLOS	UN Convention on the Law of the Sea
UNDP	UN Development Program
UNECE	United Nations Economic Commission for Europe
UNEP	UN Environment Program
UNESCO	UN Education, Scientific and Cultural Organization
UNFCCC	United Nations Framework Convention on Climate Change
UNICEF	UN Children's Fund
UNICHAL	International Union of Heat Distributors
UNIDO	UN Industrial Development Organization
UNU	United Nations University
UOD	Ultimate Oxygen Demand
UP	Ultrafine Particle
UPW	Ultrapure Water

URL	Universal Resource Locator
URV	Unit Risk Value
USC	United States Code
USCG	US Coast Guard
USCGS	US Coast and Geodetic Survey
USCSC	Ultra Supercritical Steam Cycle
USDA	US Department of Agriculture
USDW	Underground Source of Drinking Water
USGS	US Geological Survey
USP	United States Pharmacopoeia
USPHS	US Public Health Service
UST	Underground Storage Tank
UV	Ultraviolet
UVR	Ultraviolet Radiation
V	Vinyl
v/v	Volume/Volume
VA	Volatile Acid
VA	Vulnerability Assessment
VBS	Volatile Biofilm Solids
VC	Vapor Compression
VCM	Vinyl Chloride Monomer
VDI	Verein Deutscher Ingenieure
VE	Visual (Visible) Emissions
VER	Variable Energy Recovery
VFA	Volatile Fatty Acid
VFVA	Vacuum Flash Vapor Absorption
VHAP	Volatile Hazardous Air Pollutant
VITO	Flemish Institute for Technological Research
VLCC	Very Large Crude Carrier
VLF	Very Low Frequency
VLLW	Very Low Level Radioactive Waste
VO	Volatile Organics
VOC	Volatile Organic Compounds (Contaminants) (Chemicals)
VOHAP	Volatile Organic Hazardous Air Pollutant
VOST	Volatile Organic Sampling Train
VPCAR	Vapor Phase Catalytic Ammonia Removal
VPH	Volatile Petroleum Hydrocarbons
VPP	Voluntary Protection Program
VSS	Volatile Suspended Solids

VTE	Vertical Tube Evaporator
VTOC	Volatile Toxic Organic Compound
VVC	Vacuum Vapor Compression
W	Watt
WAC	Weak Acid Cation
WAMAP	IAEA Waste Management Advisory Program
WAS	Waste Activated Sludge
WASP	Water Quality Analysis and Simulation Program
WATRP	IAEA Waste Management Assessment and Technical Review Program
WBGU	German Advisory Council on Global Change
WC	Water Closet
WCD	World Commission on Dams
WCED	World Commission on Environment and Development
WCRP	World Climate Research Program
WCS	Wildlife Conservation Society
WCS	World Conservation Strategy
WDF	Waste Derived Fuel
WEAO	Water Environment Association of Ontario
WEC	Wave Energy Converter
WEC	World Energy Conference
WEC	World Energy Council
WEDO	Women and Environment Development Organization
WEF	Water Environment Federation
WEF	World Environment Federation
WEPSD	World Engineering Partnership for Sustainable Development
WES	Women's Engineering Society (UK)
WESP	Wet Electrostatic Precipitator
WFEO	World Federation of Engineering Organizations
WFI	Water for Injection (*aqua ad injectabilia*)
WFP	World Food Program
WFS	World Food Survey
WFTO	World Federation of Technology Organizations
Wh	Watt Hour
WHMIS	Waste Hazardous Materials Information System
WHO	World Health Organization
WHRC	Woods Hole Research Center
WID	Waste Incineration Directive (EU)

WIPP	Waste Isolation Pilot Plant
WIRE	World Wide Information System for Renewable Energy
WISE	Women Into Science and Engineering
WITT	Women in Trade and Technology (Canada)
WL	Working Level
WLFO	Wet Limestone Forced Oxidation
WLM	Working Level Month
WMO	World Meteorological Organization
WO	Water in Oil
WPCF	Water Pollution Control Federation
WPU	Purified Water (*aqua purifacta*)
WPWP	Western Pacific Warm Pool
WQCV	Water Quality Control Volume
WQI	Water Quality Indices
WQM	Water Quality Management
WRA	Waste Regulation Authority (UK)
WRC	World Resources Council
WRDA	Water Resources Development Act (US)
WRI	World Resources Institute
WRPA	Water Resources Planning Act (US)
WRRA	Water Resources Research Act (US)
WSC	World Solar Commission
WSI	Water-Steam Injection
WSP	World Solar Programme
WSR	World Standardized Rates
WSRA	Wild and Scenic Rivers Act (US)
WSSD	World Summit on Sustainable Development
WSSP	World Solar Summit Process
WTF	Water Treatment Facility
WTO	World Trade Organization
WWF	Wet Weather Flow
WWF	World Water Forum
WWF	World Wildlife Fund
WWI	World Watch Institute
WWTP	Wastewater Treatment Plant
WWW	World Weather Watch (WMO)
XLPE	Crosslinked Polyethylene
XRD	X-Ray Diffraction

XRF	X-Ray Fluorescence
Y	Growth Yield Coefficient
ZEV	Zero Emission Vehicle
ZIP	Zero-Incident Performance
ZLD	Zero Liquid Discharge
ZPG	Zero Population Growth

The Elements

Based on the Atomic Weight of Carbon 12 (Values given in brackets are the mass number of the longest lived or most significant isotope)

Element	Symbol	Atomic Number	Atomic Weight
Actinium	Ac	89	(227)
Aluminum	Al	13	26.98
Americium	Am	95	(243)
Antimony	Sb	51	121.75
Argon	Ar	18	39.95
Arsenic	As	33	74.92
Astatine	At	85	(210)
Barium	Ba	56	137.34
Berkelium	Bk	97	(249)
Beryllium	Be	4	9.01
Bismuth	Bi	83	208.98
Boron	B	5	10.81
Bromine	Br	35	79.91
Cadmium	Cd	48	112.40
Calcium	Ca	20	40.08
Californium	Cf	98	(251)
Carbon	C	6	12.01
Cerium	Ce	58	140.12
Cesium	Cs	55	132.91

Element	Symbol	Atomic Number	Atomic Weight
Chlorine	Cl	17	35.45
Chromium	Cr	24	52.00
Cobalt	Co	27	58.93
Copper	Cu	29	63.54
Curium	Cm	96	(247)
Dysprosium	Dy	66	162.50
Einsteinium	Es	99	(254)
Erbium	Er	68	167.26
Europium	Eu	63	151.96
Fermium	Fm	100	(253)
Fluorine	F	9	19.00
Francium	Fr	87	(223)
Gadolinium	Gd	64	157.25
Gallium	Ga	31	69.72
Germanium	Ge	32	72.59
Gold	Au	79	196.97
Hafnium	Hf	72	178.49
Helium	He	2	4.00
Holomium	Ho	67	164.93
Hydrogen	H	1	1.01
Indium	In	49	114.82
Iodine	1	53	126.91
Iridium	Ir	77	192.20
Iron	Fe	26	55.85
Krypton	Kr	36	83.80
Lanthanum	La	57	138.91
Lawrencium	Lw	103	(257)
Lead	Pb	82	207.19
Lithium	Li	3	6.94
Lutetium	Lu	71	174.97
Magnesium	Mg	12	24.31
Manganese	Mn	25	54.94
Mendelevium	Md	101	(256)
Mercury	Hg	80	200.59
Molybdenum	Mo	42	95.94
Neodymium	Nd	60	114.24
Neon	Ne	10	20.18

Element	Symbol	Atomic Number	Atomic Weight
Neptunium	Np	93	(237)
Nickel	Ni	28	58.71
Niobium	Nb	41	92.91
Nitrogen	N	7	14.01
Nobelium	No	102	(253)
Osmium	Os	76	190.20
Oxygen	O	8	16.00
Palladium	Pd	46	106.40
Phosphorus	P	15	30.97
Platinum	Pt	78	195.09
Plutonium	Pu	94	(242)
Polonium	Po	84	(210)
Potassium	K	19	39.10
Praseodymium	Pr	59	140.91
Promethium	Pm	61	(145)
Proactinium	Pa	91	(231)
Radium	Ra	88	(226)
Radon	Rn	86	(222)
Rhenium	Re	75	186.20
Rhodium	Rh	45	102.91
Rubidium	Rb	37	85.47
Ruthenium	Ru	44	101.07
Samarium	Sm	62	150.35
Scandium	Sc	21	44.96
Selenium	Se	34	78.96
Silicon	Si	14	28.09
Silver	Ag	47	107.87
Sodium	Na	11	22.99
Strontium	Sr	38	87.62
Sulfur	S	16	32.06
Tantalum	Ta	73	180.95
Technetium	Tc	43	(99)
Tellurium	Te	52	127.60
Terbium	Tb	65	158.92
Thallium	Tl	81	204.37
Thorium	Th	90	232.04
Thulium	Tm	69	168.93

Element	Symbol	Atomic Number	Atomic Weight
Tin	Sn	50	118.69
Titanium	Ti	22	47.90
Tungsten	W	74	183.85
Uranium	U	92	238.03
Vanadium	V	23	50.94
Xenon	Xe	54	131.30
Ytterbium	Yb	70	173.04
Yttrium	Y	39	88.91
Zinc	Zn	30	65.37
Zirconium	Zr	40	91.22

International System (SI) of Units

Dimension	Unit	Symbol
Length	Meter	m
Mass	Kilogram	kg
Temperature	Kelvin	K
Time	Second	s
Electric current	Ampere	A
Luminous intensity	Candela	cd
Amount of substance	Mole	mol
Derived units		
Force	Newton	N
Energy	Joule	J
Power	Watt	W
Pressure	Pascal	Pa

SI Unit Prefixes

Prefix	Symbol	Multiple
Exa	E	10^{18}
Peta	P	10^{15}
Tera	T	10^{12}
Giga	G	10^{9}
Mega	M	10^{6}
Kilo	k	10^{3}
Hecto	h	10^{2}
Deka	da	10
Deci	d	10^{-1}
Centi	c	10^{-2}
Milli	m	10^{-3}
Micro	m̈	10^{-6}
Nano	n	10^{-9}
Pico	p	10^{-12}
Femto	f	10^{-15}
Atto	a	10^{-18}

Conversion of Units

Length

1 in	= 2.54 cm	= 0.0254 m
1 ft	= 0.3048 m	
1 yd	= 0.914 m	
1 mi	= 1.609 km	

Area

1 in^2	= 6.45 cm^2	= 6.45 $\times$ 10^{-4} m^2
1 ft^2	= $^=$ 929 cm^2	= 0.0929 m^2
1 yd^2	= 8361 cm^2	= 0.8361 m^2
1 mi^2	= 2.59 æ 10^{10} cm^2	= $^=$ 2.59 $\times$ 10^6 m^2
1 ac	= 43560 ft^2	= 0.405 ha
1 hectare	= 2.47 ac	

Volume

1 in^3	= 16.39 cm^3	= 1.639 $\times$ 10^{-5} m^3
1 ft^3	= $^=$ 28317 cm^3	= $^=$ 2.83 $\times$ 10^{-2} m^3
1 gal (US)	= 3785 cm^3	= $^=$ 3.785 L
1 gal (UK)	= 4550 cm^3	= $^=$ 4.55 L

Mass

1 lb	= 453.6 g	= 0.4536 kg
1 ton	= 907185 g	= 907.19 kg

Pressure

$14.70 \text{ lb/in}^2 = 1 \text{ atm} = 29.92 \text{ in Hg} = 760 \text{ mm Hg}$

$1 \text{ Pa} \quad = 1 \text{ N/m}^2 = 10^{-5} \text{ bar} = 7.50 \times 10^{-3} \text{ mm Hg}$
$\quad\quad\quad = 10 \text{ dynes/cm}^2$

$1 \text{ bar} \quad = 10^5 \text{ Pa} = 0.9869 \text{ atm} = 750 \text{ mm Hg} = 29.50 \text{ in Hg}$
$\quad\quad\quad = 14.50 \text{ lb/in}^2$

$1 \text{ Torr} \quad = 1 \text{ mm Hg} = 133.3 \text{ Pa}$

Energy

$1 \text{ cal} \quad = 4.184 \text{ J} = 3.966 \times 10^{-3} \text{ Btu}$

$1 \text{ Btu} \quad = 1055 \text{ J} = 252 \text{ cal} = 2.93 \times 10^{-4} \text{ kw-hr} = 778 \text{ ft-lb}$
$\quad\quad\quad = 10.41 \text{ L-atm}$

$1 \text{ J} \quad = 0.239 \text{ cal} = 9.48 \times 10^{-4} \text{ Btu} = 2.778 \text{ kw-hr}$
$\quad\quad\quad = 3.725 \text{ hp-hr}$

$1 \text{ erg} \quad = 1 \times 10^{-7} \text{J}$

Unit Abbreviations

ac	acre
atm	atmosphere
Btu	British thermal unit
cal	calorie
cm	centimeter
ft	foot
g	gram
gal	gallon
ha	hectare
hp	horsepower
hr	hour
in	inch
J	joule
k	kilo
kg	kilogram
km	kilometer
kw	kilowatt
L	liter
lb	pound
m	meter
mi	mile
mm	millimeter
N	newton
Pa	pascal
Torr	torricelli
yd	yard